THE GULF RESEARCH PROGRAM

A STRATEGIC VISION

Advisory Group
Gulf Research Program

THE NATIONAL ACADEMIES PRESS
Washington, D.C.
www.nap.edu

THE NATIONAL ACADEMIES
Advisers to the Nation on Science, Engineering, and Medicine

The **National Academy of Sciences** is a private, nonprofit, self-perpetuating society of distinguished scholars engaged in scientific and engineering research, dedicated to the furtherance of science and technology and to their use for the general welfare. Upon the authority of the charter granted to it by the Congress in 1863, the Academy has a mandate that requires it to advise the federal government on scientific and technical matters. Dr. Ralph J. Cicerone is president of the National Academy of Sciences.

The **National Academy of Engineering** was established in 1964, under the charter of the National Academy of Sciences, as a parallel organization of outstanding engineers. It is autonomous in its administration and in the selection of its members, sharing with the National Academy of Sciences the responsibility for advising the federal government. The National Academy of Engineering also sponsors engineering programs aimed at meeting national needs, encourages education and research, and recognizes the superior achievements of engineers. Dr. C. D. Mote, Jr., is president of the National Academy of Engineering.

The **Institute of Medicine** was established in 1970 by the National Academy of Sciences to secure the services of eminent members of appropriate professions in the examination of policy matters pertaining to the health of the public. The Institute acts under the responsibility given to the National Academy of Sciences by its congressional charter to be an adviser to the federal government and, upon its own initiative, to identify issues of medical care, research, and education. Dr. Victor J. Dzau is president of the Institute of Medicine.

The **National Research Council** was organized by the National Academy of Sciences in 1916 to associate the broad community of science and technology with the Academy's purposes of furthering knowledge and advising the federal government. Functioning in accordance with general policies determined by the Academy, the Council has become the principal operating agency of both the National Academy of Sciences and the National Academy of Engineering in providing services to the government, the public, and the scientific and engineering communities. The Council is administered jointly by both Academies and the Institute of Medicine. Dr. Ralph J. Cicerone and Dr. C. D. Mote, Jr., are chair and vice chair, respectively, of the National Research Council.

www.national-academies.org

THE NATIONAL ACADEMIES PRESS   500 Fifth Street, NW   Washington, DC 20001

NOTICE: The project that is the subject of this document was approved by the Governing Board of the National Research Council, whose members are drawn from the councils of the National Academy of Sciences, the National Academy of Engineering, and the Institute of Medicine. The members of the Advisory Group responsible for the document were chosen for their special competences and with regard for appropriate balance.

International Standard Book Number-13: 978-0-309-31306-3
International Standard Book Number-10: 0-309-31306-6

Copies of this document can be downloaded at www.nap.edu/catalog.php?record_id=18962

Copyright 2014 by the National Academy of Sciences. All rights reserved.
Printed in the United States of America. Printed on recycled paper.

ADVISORY GROUP FOR THE GULF RESEARCH PROGRAM

BARBARA A. SCHAAL (*Chair*), Washington University, St. Louis, Missouri
DONALD F. BOESCH, University of Maryland, Cambridge
ROBERT S. CARNEY, Louisiana State University, Baton Rouge
STEPHEN R. CARPENTER, University of Wisconsin, Madison
CORTIS K. COOPER, Chevron Corporation, San Ramon, California
COURTNEY COWART, Sewanee: The University of the South, Sewanee, Tennessee
ROBERT A. DUCE, Texas A&M University, College Station
DEBORAH L. ESTRIN, Cornell New York City Tech, New York
CHRISTOPHER B. FIELD, Carnegie Institution for Science, Stanford, California
GERARDO GOLD-BOUCHOT, Center for Research and Advanced Studies at Merida, Yucatan, Mexico
LYNN R. GOLDMAN, George Washington University, Washington, DC
BERNARD D. GOLDSTEIN, University of Pittsburgh, Pennsylvania
THOMAS O. HUNTER, Sandia National Laboratories (retired), Albuquerque, New Mexico
SHIRLEY ANN JACKSON, Rensselaer Polytechnic Institute, Troy, New York
ASHANTI JOHNSON, University of Texas, Arlington, and Institute for Broadening Participation, Damariscotta, Maine
DAVID M. KARL, University of Hawaii, Honolulu
MOLLY McCAMMON, Alaska Ocean Observing System, Anchorage
LINDA A. McCAULEY, Emory University, Atlanta, Georgia
J. STEVEN PICOU, University of South Alabama, Mobile
EDUARDO SALAS, University of Central Florida, Orlando
KERRY MICHAEL ST. PÉ, Barataria-Terrebonne National Estuary Program (retired), Thibodaux, Louisiana
ARNOLD F. STANCELL, Mobil Oil (retired), Greenwich, Connecticut
LaDON SWANN, Mississippi-Alabama Sea Grant Consortium, Ocean Springs, Mississippi
JAMES W. ZIGLAR, Van Ness Feldman, Washington, DC
MARK D. ZOBACK, Stanford University, Stanford, California

Staff

CHRIS ELFRING, Executive Director
BETHANY MABEE, Research Associate
LEIGHANNE OLSEN, Senior Program Officer
JOCELYN OSHRIN, 2014 Christine Mirzayan Science & Technology Policy Graduate Fellow
EVONNE TANG, Senior Program Officer
TERI THOROWGOOD, Manager, Administrative Services
KIM WADDELL, Senior Program Officer
MAGGIE WALSER, Senior Program Officer

PREFACE 2

SUMMARY 4

1 PROGRAM FOUNDATION 11

Legal Context for Establishing the Gulf Research Program 11
First Steps in Program Planning 13
Interpreting the Settlement Language 17

2 PROGRAM MISSION, GOALS, AND OBJECTIVES 23

Mission 23
Enhance Oil System Safety 24
Improve Understanding of Human Health–Environment Connections 27
Advance Understanding of the Gulf of Mexico as a Dynamic System 29
Program Objectives 31

3 STRATEGIES FOR LASTING BENEFIT 33

Long-Term, Cross-Boundary Perspective 33
Science to Advance Understanding 34
Science to Serve Community Needs 35
Synthesis and Integration 36
Coordination and Partnerships 38
Leadership and Capacity Building 39

4 INITIAL AND FUTURE ACTIVITIES 41

Initial 2015–2016 Activities 41
Next Steps for 2015–2020 45
Next Steps for Identifying Long-Term Initiatives 46

5 PROGRAM OPERATIONS 49

Scientific Integrity 49
Coordination and Communication 50
Planning and Evaluation 51

REFERENCES 55

APPENDIXES 57

A Advisory Group Biographies 57
B Charge to the Advisory Group 63
C Outreach Activities 64
D Other Funding Programs 66

PREFACE

In 2013 the U.S. Department of Justice asked the National Academy of Sciences (NAS) to accept funds from the settlement of the federal criminal complaints against BP Exploration & Production Inc. and Transocean Deepwater Inc. as a result of the 2010 *Deepwater Horizon* explosion and fire, and use the funds to establish a new program focused on oil system safety, environmental resources, and human health in the Gulf of Mexico and other regions on the U.S. outer continental shelf. The program is to be entirely independent, with the companies having no role in directing use of the funds. The funds are being received between 2013 and 2018 and are to be expended over 30 years.

This opportunity to direct some of the settlement funds toward scientific activities that benefit the Gulf region and the Nation is not taken lightly. The program can help advance scientific understanding of this important region and enhance the Nation's capabilities to prevent, plan for, mitigate, and respond to future disasters. It is also an opportunity to think broadly about the connections among energy production, the environment, and health. Thus, in summer 2013, the NAS and its affiliated organizations—the National Academy of Engineering (NAE), the Institute of Medicine (IOM), and the National Research Council (NRC)—together embarked on a planning process to shape the new program.

An Advisory Group (see Appendix A) with extensive expertise was appointed to guide the creation of the Gulf Research Program. It was asked to establish the foundation of a Program that, over time, is expected to make an enduring contribution to the vitality of the Gulf region. The Advisory Group set out to determine how the Program could build relationships at the state and federal levels; translate the broad mandates of the settlement agreements into a vision of the Program's mission, goals, and objectives; and propose an initial set of activities.

We want to thank the members of the Advisory Group for their dedicated service—they have been deeply engaged and insightful. The chair, Dr. Barbara Schaal, Dean of the Faculty of Arts and Sciences at Washington University, St. Louis, merits special recognition for leading this energetic group of 25 volunteers. We also want to thank the nearly 300 people who offered their advice and perspectives during the planning process. We look forward to a continuing conversation—with government agencies; nonprofit associations; scientists, engineers, and physicians; industries; and citizens—to ensure that the Program fulfills its intended purpose.

This document presents a vision to guide the new Gulf Research Program, but much work remains to be done to turn this foundation into an effective suite of work. In fall 2014, oversight for the Gulf Research Program will transition to a newly appointed Advisory Board charged to implement the Program's vision, and we look forward to the new board's contributions.

We recognize that the Gulf Research Program will evolve over its 30-year duration, but we believe that this document provides a strong foundation. The NAS, NAE, IOM, and NRC will work together to ensure the Program's success.

RALPH J. CICERONE, President, National Academy of Sciences
C. D. MOTE, Jr., President, National Academy of Engineering
VICTOR J. DZAU, President, Institute of Medicine

SUMMARY

On April 20, 2010, an explosion and subsequent fire on the *Deepwater Horizon* oil drilling platform in the Gulf of Mexico killed 11 workers and injured 17 others. After burning for more than 1 day, the *Deepwater Horizon* sank in approximately 5,000 feet of water. For the next 87 days, the well that the *Deepwater Horizon* had been drilling released oil into the Gulf, resulting in the largest offshore oil spill in U.S. history and causing serious economic, environmental, and health hazards and hardships for the Gulf region.

As part of agreements settling criminal charges against the companies held responsible for the spill—BP Exploration & Production Inc. (BP) and Transocean Deepwater Inc. (Transocean)— the Department of Justice asked the National Academy of Sciences (NAS) to establish a new program focused on oil system safety, environmental resources, and human health in the Gulf of Mexico and other U.S. outer continental shelf regions that support oil and gas production. This program, known as the Gulf Research Program, is to be supported by $500 million paid by BP and Transocean between 2013 and 2018, with the funds to be expended over 30 years, until 2043.

Given its $500 million endowment and 30-year duration, the Gulf Research Program presents an extraordinary opportunity to tackle large, complex issues at a regional scale and over the long term. It has the potential to have significant impacts.

Beginning in summer 2013, an independent Advisory Group was tasked to develop a strategic vision to guide the Gulf Research Program. The Advisory Group worked for 1 year, gathering input from individuals and organizations in the Gulf region, learning from other organizations with similar missions, and identifying needs that align with the Program's specified mandate. The Group also planned a small number of initial, short-term activities to begin in 2014 even as further Program development continues. This document describes the Gulf Research Program's initial focus and is expected to guide its work over its first 5 years (2015–2020), but with the recognition that the Program will evolve over time.

Mission of the Gulf Research Program

Over its 30-year duration, the Gulf Research Program will work to enhance oil system safety and the protection of human health and the environment in the Gulf of Mexico and other U.S. outer continental shelf areas by seeking to improve understanding of the region's interconnecting human, environmental, and energy systems and fostering application of these insights to benefit Gulf communities, ecosystems, and the Nation.

Program Goals

The Gulf Research Program's most valuable contributions are likely to come at the intersections of its areas of responsibility—oil system safety, human health, and environmental resources. Given this context, the Program will address three interconnected goals:

Goal 1: Foster innovative improvements to safety technologies, safety culture, and environmental protection systems associated with offshore oil and gas development.

Goal 2: Improve understanding of the connections between human health and the environment to support the development of healthy and resilient Gulf communities.

Goal 3: Advance understanding of the Gulf of Mexico region as a dynamic system with complex, interconnecting human and environmental systems, functions, and processes to inform the protection and restoration of ecosystem services.

Program Objectives

To support these goals in the first 5 years (2015–2020), the Program will pursue the following broad objectives through a variety of activities and approaches:

- Partner with industry, government, and academia to identify key opportunities for enhancing the safety of offshore energy development.

- Explore models of decision-support systems for safe and environmentally sustainable offshore oil and gas development, disaster response, and remediation options.

- Provide research opportunities that improve understanding of how social, economic, and environmental factors influence community vulnerability, recovery, and resilience.

- Support research, long-term observations and monitoring, and information development to advance understanding of environmental conditions, ecosystem services, and community health and well-being in the Gulf of Mexico.

- Support the development of future professionals and leaders—in science, industry, health, policy, and education—who apply cross-boundary approaches to critical issues that span oil system safety, human health, and environmental resources.

- Identify opportunities for knowledge transfer between the Gulf of Mexico and other U.S. outer continental shelf regions.

- Support activities to improve understanding and use of scientific information by the public and policy makers in decisions related to environmental stewardship, human health improvement, and responsible oil and gas production.

As specified in the agreements that established the Program, three broad approaches, or mechanisms, will be used: research and development, education and training, and environmental monitoring and assessment.

Strategies to Achieve Lasting Benefit

The planning process identified six overarching strategies that can steer the Program toward producing lasting benefit. These are key opportunities where the mission of the Gulf Research Program aligns with the strengths of the National Academies and where the 30-year duration and long-term perspective hold special potential for cumulative impacts.

Long-Term, Cross-Boundary Perspective. Two distinctive features of the Program are the 30-year duration and the geographic focus that extends beyond the Gulf of Mexico to also include other U.S. outer continental shelf regions. The Program will attempt to select areas of work that take advantage of a long-term perspective and result in the transfer of knowledge between the Gulf of Mexico and other offshore energy-producing regions. The Program will also encourage work across state, disciplinary, and sectoral boundaries.

Science to Advance Understanding. A fundamental purpose of the Program is to bring the best expertise in science, engineering, technology, and health to advance understanding of the Gulf region in the context of linkages among people, ecosystems, and energy development. The Program aims to encourage innovative thinking and approaches and potentially transformative science and technology.

Science to Serve Community Needs. The Program seeks to foster science that serves the needs of the region's numerous and diverse communities, including translational research that is focused on the ways in which new knowledge can be used by the public, resource managers, program managers, community planners, and other decision makers.

Synthesis and Integration. Given the amount of data and information already available about the Gulf of Mexico, the Program envisions significant opportunities in the synthesis and integration of data and information, especially across disciplines, to produce novel insights and accelerate the translation of new understanding into action.

Coordination and Partnerships. Being one program among many operating in the Gulf region, the Program recognizes the importance of coordination to avoid duplication and leverage resources. The 150-year history of the NAS as an independent, nonprofit organization devoted to consensus building positions the Program to provide leadership and participate in efforts to facilitate coordination and build partnerships among the many groups and organizations operating in the Gulf region.

Leadership and Capacity Building. By investing in leadership and capacity building, the Program hopes to provide opportunities for academic and community leaders, state and regional decision makers, students, and institutions to develop skills, competencies, and capabilities that are needed to solve problems, spark innovation, and establish sustainable systems, economies, and communities.

Initial and Future Activities

A suite of initial, short-term activities will be announced in 2014 and funded in 2015, even while planning for larger and longer-term activities continues. The first calls for applications will be in three areas: exploratory grants, research fellowships, and science policy fellowships. A fourth opportunity related to synthesis and integration of environmental monitoring data will be offered in early 2015. The Program also expects to support expert consensus studies of value to the Gulf region, planning meetings to inform the Program's future activities, and workshops and other mission-relevant activities.

As these initial activities get under way, in fall 2014 a new Advisory Board will take over Program development and oversight and will be charged to identify larger and more far-reaching themes and activities to achieve the Program's goals and objectives. In 2015 the Program will use the results of three opportunity analysis workshops to begin additional activities, in the areas of training middle-skilled workers, community resilience and health, and environmental monitoring. It will begin exploratory discussions to identify a few larger, long-term activities. The incoming Advisory Board will work to determine future potential activities that address the program's mission and objectives, align with the strengths of the National Academies, and increase the Program's impact.

EXPLORATORY GRANTS

2015 Topics:
- Exploring approaches for effective education and training of workers in the offshore oil and gas industry and health professions
- Linking ecosystem services related to and influenced by oil and gas production to human health and well-being

Expected 2016 Topics:
- Innovative approaches to developing scenario planning and decision-support systems to cope with crises
- Connecting data about environmental conditions with individual and population health data to foster transdisciplinary research
- Building resilience in human and environmental systems of the Gulf of Mexico and other offshore energy-producing regions

FELLOWSHIPS

Early-Career Research Fellowships:
Two-year fellowships for pretenure faculty, recognizing exceptional leadership, past performance, and potential for future contributions to improving oil system safety, the environment, or human health.

Science Policy Fellowships:
One-year fellowships that will contribute to leadership development and capacity building by providing recipients with a valuable educational experience at the science-policy interface.

Christine Mirzayan Science & Technology Policy Graduate Fellowship:
An existing National Academies fellowship that introduces early-career professionals to the role of science in the federal government. One fellow will be hosted each year by the Gulf Research Program for this 12-week opportunity in Washington, DC.

ENVIRONMENTAL MONITORING

Integration and Synthesis of Monitoring Data Opportunity:
Applicants will be challenged to propose hypothesis-driven projects that identify and synthesize existing data related to either the deep Gulf or ecosystem services for restoration and management themes.

WORKSHOPS

2014:
Education and Training Opportunity Analysis Workshop
Environmental Monitoring Opportunity Analysis Workshop
Community Resilience and Health Opportunity Analysis Workshop

2015:
Additional workshops to be determined.

1
PROGRAM FOUNDATION

Legal Context for Establishing the Gulf Research Program   11

First Steps in Program Planning   13

Interpreting the Settlement Language   17

The Gulf of Mexico region is home to diverse and vibrant communities, productive ecosystems, and thriving industries. It constitutes an invaluable part of the Nation's population, economy, and natural resources. The 2010 Macondo Well *Deepwater Horizon* (*DWH*) oil spill placed a spotlight on the importance of this region and on the deep connections between Gulf residents and their surrounding ecosystems.

The *DWH* oil spill[1] began with an explosion and fire that led to the deaths of 11 oil rig workers and the injury of 17 others. For 87 days after the *DWH* rig sank, oil flowed from a blowout of the Macondo well, causing the largest offshore oil spill on record in U.S. waters. In total, approximately 172 million gallons of oil spilled (McNutt et al., 2012), causing significant harm to the Gulf of Mexico environment and its people and testing the resilience of communities still struggling to recover from the devastating effects of a series of hurricanes in 2005 and 2008.

The recovery of the Gulf region from the disaster has been arduous and is ongoing. But the recovery and restoration process also provides an opportunity to help communities strengthen their capacity to meet future challenges.

Legal Context for Establishing the Gulf Research Program

As part of agreements settling the criminal charges against BP Exploration & Production Inc. (BP) and Transocean Deepwater Inc. (Transocean), the companies responsible for the *DWH* oil spill, the U.S. Department of Justice asked the National Academy of Sciences (NAS) to establish a new program focused on oil system safety, human health, and environmental resources in the Gulf of Mexico and the U.S. outer continental shelf. In aggregate, the NAS will receive $500 million between 2013 and 2018 for the Gulf Research Program, which will be placed in an endowment and expended within 30 years (see Box on p. 12). The companies are prohibited from having any role in the Program.

The agreements provide broad guidance on how the funds are to be used. The Program is expected to consist of studies, projects, and other activities using three approaches—research and development, education and training, and environmental monitoring. The Program will engage the nation's scientific, engineering, and health communities with the overarching objectives

---

[1] References to the "*DWH* oil spill" throughout this document include the related events of the Macondo well blowout, the destruction and sinking of the *Deepwater Horizon* oil rig, and the subsequent oil spill.

FUNDING SCHEDULE

As part of the agreements settling criminal charges against the two companies held responsible for the *DWH* oil spill (BP Exploration & Production Inc. and Transocean Deepwater Inc.), the Department of Justice asked the National Academy of Sciences (NAS) to establish a $500-million, 30-year program with the objectives of enhancing oil system safety and improving the protection of human health and environmental resources. Under the agreements, BP will pay $350 million by 2018 and Transocean will pay $150 million by 2017 to the NAS to be managed in a fixed-term endowment.

	BP payments	Transocean payments
2013	$5 million	$2 million
2014	$15 million	$7 million
2015	$45 million	$21 million
2016	$80 million	$60 million
2017	$90 million	$60 million
2018	$115 million	

of "enhancing the safety of offshore oil drilling and hydrocarbon production" and protecting "human health and environmental resources in the Gulf of Mexico and United States outer continental shelf." The program's work is to be carried out in the public interest, supporting activities that otherwise might not be pursued.

The agreements require that at least once per year the Program will interact with the environmental protection departments and other natural resource managers in the Gulf States (Alabama, Florida, Louisiana, Mississippi, and Texas) and at the federal level with the Interagency Coordinating Committee on Oil Pollution Research (ICCOPR)[2] and its participating members, particularly the Department of the Interior's Bureau of Safety and Environmental Enforcement (BSEE) and Bureau of Ocean Energy Management (BOEM). Beyond this broad guidance, the Program is to be independently planned and implemented by the NAS.

---

[2]ICCOPR comprises 15 members representing federal independent agencies, departments, and department components: Bureau of Ocean Energy Management (BOEM), Bureau of Safety and Environmental Enforcement (BSEE), Department of Energy, Environmental Protection Agency (EPA), Federal Emergency Management Agency (FEMA), Maritime Administration, National Aeronautics and Space Administration, National Institute of Standards and Technology, National Oceanic and Atmospheric Administration (NOAA), Pipeline and Hazardous Materials Safety Administration, U.S. Army Corps of Engineers, U.S. Coast Guard (USCG), U.S. Fire Administration, U.S. Fish and Wildlife Service, and U.S. Navy. USCG chairs the ICCOPR; NOAA, BSEE, and EPA rotate assignments as vice chair every 2 years.

First Steps in Program Planning

The NAS was chosen to lead a new program[3] because of its long history as an independent, objective adviser to the Nation in areas of science, engineering, technology, and health. As a private, nonprofit organization with a record of 150 years of public service, the NAS brings an independent view and objective leadership to the design, implementation, and oversight of the Program.

Creating a new program of this scale and complexity requires careful planning. To begin, the NAS established an expert group of volunteers from diverse backgrounds to guide the planning process and develop an initial, strategic foundation for the Program. The Advisory Group began its work in summer 2013 to articulate the initial program design and focus (see Appendixes A and B). The group first sought answers to several basic questions: What are the Program's assigned responsibilities? What are other funding bodies doing that potentially interact or overlap? What are some of the perceived needs that align with the Program's responsibilities and the strengths of the Academies? What initial activities should be undertaken?

Significant effort was made to understand the context in which the Program will operate. The Advisory Group and Program staff explored the meaning of the tasks specified in the legal agreements, built relationships with other organizations, and identified perceived needs that match the Program's mandate. To understand the existing landscape, the Advisory Group convened a series of meetings in the Gulf States along with virtual meetings to learn from and interact with a wide range of stakeholders. Through these meetings, participation in conferences, and other discussions (see Appendix C), the Advisory Group learned from many people familiar with past and

THE NATIONAL ACADEMY OF SCIENCES

To meet the government's need for an independent adviser on scientific matters, President Lincoln signed a congressional charter in 1863 forming the National Academy of Sciences (NAS) to "investigate, examine, experiment, and report upon any subject of science." As science came to play an ever-increasing role in national priorities and public life, the NAS, a private, nonprofit organization, expanded to include the National Research Council (NRC) in 1916, the National Academy of Engineering (NAE) in 1964, and the Institute of Medicine (IOM) in 1970.

These organizations, together known as the National Academies, have a mission to increase public understanding of science, foster the wise use of science in decision making and public policy, and promote the acquisition and dissemination of knowledge in matters involving science, engineering, technology, and health. The institution is widely respected for its independence and has unique capabilities to bring the best science to bear on issues of national importance. In any given year, nearly 7,000 experts volunteer their time and expertise to work on Academies projects, including a number with direct relevance to the Gulf Research Program (e.g., *Macondo Well Deepwater Horizon Blowout: Lessons for Improving Offshore Drilling* [NAE and NRC, 2011] and *Assessing the Effects of the Gulf of Mexico Oil Spill on Human Health* [IOM, 2010]).

Activities at the National Academies include consensus studies; convening activities such as workshops, forums, roundtables, and symposia; management of science program reviews, fellowships, and other educational activities; and publication of the *Proceedings of the National Academy of Sciences of the United States of America*. The NAS, NAE, IOM, and NRC do not receive direct appropriations from the federal government, although specific activities often are funded by federal agencies. Other sources of funds include foundations, state governments, and the private sector.

---

[3]Two other nonprofit organizations, the North American Wetlands Conservation Fund (NAWCF) and the National Fish and Wildlife Foundation (NFWF), also received funds arising out of the criminal settlements. NAWCF is receiving $100 million to be used for wetlands conservation projects. NFWF is receiving $2.544 billion to be used for natural resources restoration in the Gulf of Mexico.

future challenges in the Gulf region, interacted with other organizations that have been funding programs related to the *DWH* oil spill, and began discussions with some of the other entities planning to conduct or fund restoration-related work in the Gulf (see Appendix D).

Through these discussions, it became clear that a substantial number of organizations and people have been doing a diverse array of related work in the Gulf region, many with years of experience. The Program wants to benefit from the experiences of others, build on existing work, and seek partnership opportunities, and is committed to continuing dialog with the many other stakeholders interested in restoring and enhancing the resilience of the Gulf region ecosystems and communities.

The Gulf of Mexico

Outreach activities helped the Advisory Group deepen its understanding of the strengths and challenges in the region and how the new Program might contribute. One significant change over time has been coastal population growth: the region's coastal counties are now home to 40 million residents and are home to 7 of the 10 busiest ports in the United States. The region is economically and culturally vibrant, but there are significant challenges due to competing uses of valuable environmental resources.

For example, the Mississippi River watershed delivers millions of metric tons of sediments from the Midwest that historically helped build and sustain coastal wetlands that, in turn, supported fisheries and wildlife. However, these sediments also can impede shipping, an industry that moves many billions of dollars' worth of commodities in and out of the country. Construction of levees, closures of tributaries, and dredging to control floods have supported the development of shipping channels and are also important to the production and transportation of oil and gas. However, these actions have also

contributed to the loss of wetlands and their associated services. Other environmental stressors, such as sea-level rise and other effects of climate change, will only exacerbate the difficulty of managing competing interests. These challenges will test the substantial resilience of the region's communities and natural ecosystems.

Few regions have such a remarkable blend of cultures, ethnicities, and histories as that found in the five U.S. states that border the Gulf of Mexico. Together, they exhibit a rich heritage—one that reflects elements of African American, Cajun, Creole, Croatian, Latino, Native American, and Vietnamese cultures, among others. This diversity continues to enrich the region culturally and economically. For example, many of these communities have multigenerational ties to industries that rely heavily on natural resources, including nature tourism, the shellfish and finfish industries, and the oil and gas industry (LSU AgCenter, 2012).

The region's abundant natural resources continue to drive economies that attract new residents and visitors. Covering 600,000 square miles, the Gulf of Mexico contains commercial and recreational fisheries that annually harvest more than 1 billion pounds of finfish and shellfish valued in excess of $600 million. Recreational fishing alone is a $2 billion per year industry. Tourism is a $20 billion per year industry that employs more than 600,000 people. The coastline of the five Gulf States is more than 3,500 miles long and borders 5 million acres of wetlands, representing more than 40 percent of the coastal wetlands in the contiguous 48 states. These wetlands provide essential habitat for the region's fisheries and for 75 percent of the Nation's migrating birds. The region also supports a trillion-dollar gas and oil industry. The Gulf States produce nearly half the Nation's natural gas and 23 percent of the Nation's oil. The offshore oil sector employs 55,000 workers.

THE CLIMATE CHANGE CONTEXT

Many lines of evidence indicate that humans are changing Earth's climate. The atmosphere and oceans have warmed, accompanied by sea-level rise, a strong decline in Arctic sea ice, and other climate-related changes. Numerous reports coming from the Intergovernmental Panel on Climate Change (IPCC, 2013, 2014a,b,c), the Third National Climate Assessment (Ingram et al., 2013), and the NAS (NRC, 2010, 2011; National Academy of Sciences and Royal Society, 2014) have drawn increasingly confident conclusions regarding the human influence on the changing global climate, current and anticipated impacts, and possible ways to limit and adapt to change.

During the 30-year duration of the Program, the consequences of changing climate will become more manifest along the Gulf Coast. Adaptation will be essential to social resilience. The fossil fuel–based economy of the Gulf region will change in ways that are difficult to foresee as the Nation seeks ways to reduce greenhouse gas emissions.

The U.S. Gulf Coast is particularly vulnerable to sea-level rise. Globally averaged sea level, which had been relatively stable over 2,000 years, rose 1.8 mm/year during the 20th century, has been rising 3.2 mm/year over the past 20 years, and with high certainty is projected to rise at greater rates during the 21st century. To compound the challenge resulting from sea-level rise, parts of the Louisiana and Texas coast are subsiding faster than the ocean is presently rising. As a result, sea level is already rising over 9 mm/year relative to the shore in the outer Mississippi River delta and nearly 7 mm/year around Galveston Bay (http://tidesandcurrents.noaa.gov/sltrends).

Along the Gulf Coast, accelerated sea-level rise will have significant effects on barrier shorelines; deltaic and other coastal wetlands; urban as well as less densely populated areas; levees and floodwalls that protect people and property, ports, oil and gas infrastructure; and large areas of coastal landscape that today lie less than 1 meter above mean sea level.

Another key vulnerability is severe weather. The Gulf Coast is the part of the United States that experiences the greatest frequency of hurricane landfalls, as well as the most intense hurricanes. The devastation caused by Katrina and other recent hurricanes highlighted the vulnerability to hurricanes of coastal environments, human populations,

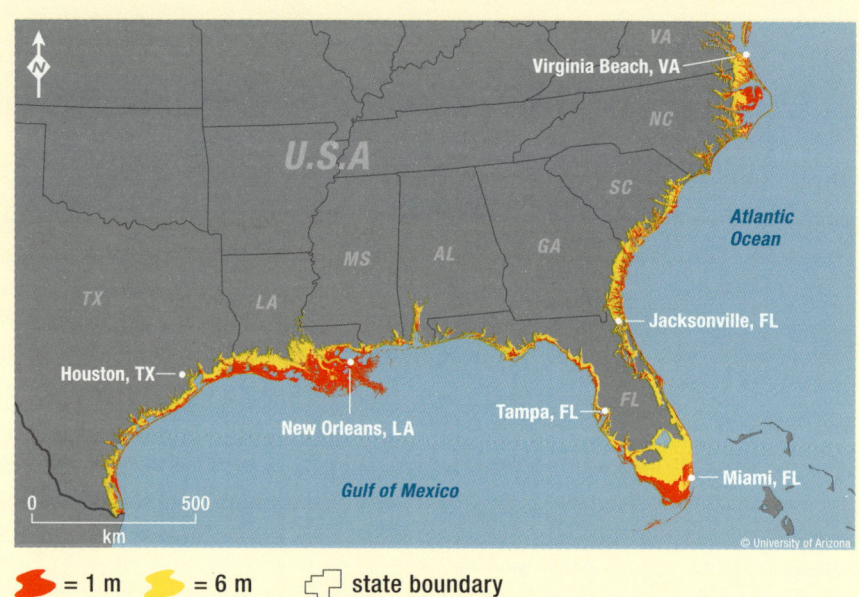

This map shows where increases in sea level could affect the southern and Gulf coasts of the United States. The colors indicate areas along the coast that are elevations of 1 meter or less (red) or 6 meters or less (yellow) and have connectivity to the sea. Credit: Weiss and Overpeck, University of Arizona. SOURCE: Weiss et al. (2011).

infrastructure, offshore energy production, and the Nation's energy supply. The dynamics of tropical cyclones are expected to be affected by global climate change; current scientific understanding indicates that although Atlantic hurricanes might not become more frequent, they are likely to become more powerful.

Other climate change impacts, including warming of the atmosphere and Gulf waters, increasing acidity of Gulf waters, and changes in precipitation, river flows, and ocean currents, also will affect the Gulf Coast. Climate models indicate an increased frequency of extremely warm days, more extreme rainfall events, and, at least along the central and western Gulf Coast, drier summer conditions. Increased precipitation in the upper Mississippi River Basin might generate greater discharges of the Mississippi and Atchafalaya Rivers, which will influence circulation, productivity, and the occurrence and extent of hypoxia (creating what is known as the Dead Zone) in the northern Gulf.

Despite these strengths and resources, the Gulf region faces a suite of challenges that vary from global to local in scale and impact. As identified in one assessment (Mabus, 2010) of the Gulf region's environment, economy, and health, these challenges include wetland loss, which in Louisiana alone averages 25 to 35 square miles per year; erosion of coastal barrier islands and shorelines, which undermines capacity for storm protection in a region that has experienced more than 120 hurricanes over the past century; loss and degradation of coastal estuarine habitats, which threaten the critical nursery habitat for the Gulf's fishery resources; overfishing of major commercially and recreationally important fish stocks; and hypoxia caused by excess nutrient pollution from the Mississippi River watershed.

Like the rest of the Nation, some of the greatest challenges to human health in the Gulf region are related to chronic illness, health disparities, and growing inequality. Environmental stressors related to climate change and the loss of ecosystem services are expected to create new health challenges for the region.

The combination of these challenges is specific to the Gulf region, but they are exacerbated by global phenomena such as coastal population increase and climate change. Over time, global responses to reduce greenhouse gas emissions are expected to have implications for the oil and gas industry, the predominant driver of the Gulf's economy.

These challenges are not unlike those faced by other regions in the United States and across the globe. The funds stemming from the *DWH* oil spill provide a remarkable opportunity for scientists, policy makers, and communities to build on the region's considerable strengths to prepare and meet future challenges.

Interpreting the Settlement Language

Given its $500 million endowment and 30-year duration, the Program presents an extraordinary opportunity to tackle large, complex issues at a regional scale and from a long-term perspective. The settlement language provides broad guidance on how the Program should contribute by identifying three areas of responsibility: oil system safety, human health, and environmental resources. The agreements also identify three broad categories of activity: research and development, education and training, and environmental monitoring.

The agreement is also clear that while the work is to focus on the Gulf of Mexico, it is to be relevant to the Nation as a whole, considering insights learned from or useful to other U.S. outer continental shelf regions where oil and gas development is occurring or being considered. In these regions, human communities, ecosystems, and oil and gas exploration and development interact. The following section expands on the broad categories of activity provided in the settlement language and defines terms to be used by the Program.

Categories of Activity

The Program will emphasize work at the intersections of its three specified areas of responsibility—oil system safety, human health, and environmental resources. For the purposes of the Program, human health is defined as "the complete state of physical, mental, and social well-being and not merely the absence of disease or infirmity."[4] The protection of environmental resources is defined as the protection and restoration of the environment and ecosystem services. Ecosystem services are the benefits that people obtain from ecosystems, including provisioning services such as the supply of food and water; regulating services such as flood and disease control; cultural services such as spiritual, recreational, and cultural benefits; and supporting services, such as nutrient cycling, that maintain the conditions for life on Earth.

The Gulf Research Program will focus on issues at the intersections of its three specified areas of responsibility.

---

[4]Preamble to the Constitution of the World Health Organization as adopted by the International Health Conference, New York, 19-22 June 1946; signed on 22 July 1946 by the representatives of 61 States (Official Records of the World Health Organization, no. 2, p. 100) and entered into force on 7 April 1948.

Research and Development

The Program will support research, technology development, information synthesis, and other science-based activities that improve oil system safety and enhance the protection of environmental resources and human health in the Gulf of Mexico and other regions that support oil and gas production on the U.S. outer continental shelf. Through research and development (R&D) activities, the Program will seek to increase understanding of the Gulf region as a dynamic system that provides key ecosystem services such as energy, seafood, and wetlands for storm protection. R&D will also be used to advance oil system safety, primarily through work on prevention and safety culture and to expand understanding of the factors that influence community resilience to disasters and future environmental change.

Some Program R&D funding will emphasize synthesis and assessment of information, which is critical to translating and packaging data in ways that can be readily incorporated into decision making by target audiences. The Program aims to fund innovative and collaborative research and related activities. The Program will seek a balance of short-, medium-, and long-term efforts, with an emphasis on cross-boundary approaches that catalyze work across disciplines, geographic borders, perspectives, and sectors. Ideally, the Program will seek a balance of risk and return, at times supporting lower-risk approaches and at times seeking innovation even if at greater risk.

Education and Training

The Program will foster capacity building and the engagement, education, and training of scientists, engineers, health professionals, and offshore oil and gas industry personnel. In its education and training (E&T) components, the Program will stress leadership development and cross-boundary thinking, and strive to engage a range of program participants that reflect the diversity of the communities where offshore oil exploration exists. The Program will support the development of tools, programs, and partnerships. It may also support the development of novel teaching and learning technologies.

The Program will avoid duplicating existing E&T programs and will seek to transfer successful models between the Gulf and other regions. The Program will support collaborative activities that span multiple disciplines and sectors. The Program will provide incentives for R&D activities that include student participation at the undergraduate and graduate levels as well as postdoctoral opportunities. These types of investments will grow and evolve over the course of the Program's 30 years and will help build future generations of science, engineering, and health professionals, and a workforce with a regional and interdisciplinary perspective.

Environmental Monitoring

For the purposes of the Program, environmental monitoring is defined as "a continuing program of measurement, analysis, and synthesis to identify and quantify ecosystem conditions and trends to provide a technical basis for decision making" (GoMRI, 2014). Monitoring information can be used to increase basic understanding, identify emerging problems and long-term trends, inform restoration projects, prioritize use of resources, and provide information to guide policy and management. It is essential for increasing understanding of how an ecosystem and its components change over time. For rapidly changing regions such as the Gulf of Mexico, monitoring efforts can yield reference data that alert stakeholders to emerging environmental and health concerns or serve as base data for a future disaster.

ENVIRONMENTAL MONITORING IN THE GULF OF MEXICO: Developing an Integrated System

Environmental monitoring programs are conducted for a number of purposes and can vary significantly in the scale of their spatial and temporal boundaries. They can also vary significantly in scope, ranging from community-based monitoring on a local scale, to large-scale collaborative global monitoring programs. In a region as large and ecologically diverse as the Gulf of Mexico, and where numerous federal, state, and local jurisdictions have differing responsibilities, it is not surprising that the region is home to a large number of environmental monitoring programs with a diversity of goals and metrics.

This diversity, when viewed at the large regional or ecosystem level, has occasionally and unintentionally created duplication of efforts and reduced the potential for integrating and synthesizing the data from different programs; this limits the usefulness of the data. Consequently, there have been increasing calls for more coordination among the programs engaged in monitoring activities around the Gulf region.

Several regional organizations had proposed elements of a collective or network approach to environmental monitoring in the Gulf, including the Gulf of Mexico Alliance (GOMA) and the Gulf of Mexico Coastal Ocean Observing System (GCOOS). Interest in increased collaboration has accelerated since the *DWH* oil spill, which because of its size, highlighted

During the Program's 30-year duration, coastal populations will continue to grow, with increased demands on coastal and marine ecosystem services. There is the possibility of future oil spills and the challenges of understanding the extent of risk (such as whether seafood is safe) and resulting damage (which requires a strong foundation of data and information before the event). The Program will operate in a context that includes other, significant environmental challenges, including climate change, sea-level rise, and coastal subsidence.

Environmental monitoring is a major, multifaceted challenge, and it is expensive. A broad partnership will need to coalesce to identify user needs, produce coordinated efforts, and identify cutting-edge technologies. The Program could serve as a catalyst in this process, seeking to leverage funds and contribute long-term and regional perspectives. It will participate in continuing community dialogs seeking the best approaches and partnerships to meet monitoring needs.

the need for a stronger and more holistic understanding of the Gulf ecosystem's health and function. The *DWH* oil spill also triggered a considerable increase in funding to investigate and assess the impacts from the spill, and these efforts are generating volumes of new environmental data and information.

For example, in summer 2010 BP committed $500 million over a 10-year period to create the Gulf of Mexico Research Initiative (GoMRI), a broad, independent research program that funds research across multiple research institutions primarily in the Gulf States. Although GoMRI's mission is primarily to investigate the impacts of the *DWH* oil spill, it also shares with GCOOS, GOMA, the Program, and many other organizations the common goal of improving the long-term environmental health of the Gulf of Mexico. All of these organizations recognize the need for an integrated ecosystem observing system for the Gulf of Mexico.

A number of organizations and agencies have come together to discuss improving the coordination of environmental monitoring efforts in the Gulf of Mexico. In GoMRI's 2014 Oil Spill Conference, a daylong session titled "Current and Future Ecosystem Monitoring Strategies in the Gulf of Mexico" was held. The session had three objectives: (1) to begin to synthesize current monitoring projects and assets in the Gulf, (2) to determine the community's observing priorities in a variety of disciplines, and (3) to elicit participant recommendations for an optimized, integrated observing system. GCOOS, in particular, is committed to implementing these recommendations by integrating them with its current and planned activities.

Other community dialogs about coordination continue. For instance, the U.S. National Oceanic and Atmospheric Administration (NOAA) National Coastal Data Development Center has worked with GOMA to organize two workshops that brought stakeholders from state and federal governments, nongovernmental organizations, industry, and academia to identify and develop priorities for observation and monitoring in the Gulf. One of the overarching needs identified at the workshops is to develop a mechanism that recovers data from past monitoring programs, stores data from current monitoring activities, and provides public-ready access to the data and related information. The workshops identified outcomes around three themes: (1) enhancing partnerships and collaboration among the agencies and entities involved in monitoring; (2) identifying existing data gaps, developing comprehensive data inventories for the region, and providing a mechanism for improved data management and infrastructure; and (3) expanding communication efforts with the public and funding agencies, both to help establish the value of the various products that come from environmental monitoring and to support science-based resource management decisions. Collectively, these community discussions are steps toward the realization of an integrated ecosystem observing system that both helps advance basic understanding of the Gulf of Mexico and fosters use of that information for decision making.

2
PROGRAM MISSION, GOALS, AND OBJECTIVES

Mission 23

Enhance Oil System Safety 24

Improve Understanding of Human Health–Environment Connections 27

Advance Understanding of the Gulf of Mexico as a Dynamic System 29

Program Objectives 31

In keeping with the spirit of the agreements negotiated by the Department of Justice, community input, and the deliberations of the Advisory Group, the Program will pursue the following mission:

Mission

Over its 30-year duration, the Gulf Research Program will work to enhance oil system safety and the protection of human health and the environment in the Gulf of Mexico and other U.S. outer continental shelf areas by seeking to improve understanding of the region's interconnecting human, environmental, and energy systems and fostering application of these insights to benefit Gulf communities, ecosystems, and the Nation.

The Program's most valuable contributions are likely to come at the intersections of its areas of responsibility: oil system safety, human health, environmental resources. Given this context, the Program will address three connected goals:

GOAL 1 Foster innovative improvements to safety technologies, safety culture, and environmental protection systems associated with offshore oil and gas development.

GOAL 2 Improve understanding of the connections between human health and the environment to support the development of healthy and resilient Gulf communities.

GOAL 3 Advance understanding of the Gulf of Mexico region as a dynamic system with complex, interconnecting human and environmental systems, functions, and processes to inform the protection and restoration of ecosystem services.

MISSION

Over its 30-year duration, the Gulf Research Program will work to enhance oil system safety and the protection of human health and the environment in the Gulf of Mexico and other U.S. outer continental shelf areas by seeking to improve understanding of the region's interconnecting human, environmental, and energy systems and fostering application of these insights to benefit Gulf communities, ecosystems, and the Nation.

GOALS

1 Foster innovative improvements to safety technologies, safety culture, and environmental protection systems associated with offshore oil and gas development.

2 Improve understanding of the connections between human health and the environment to support the development of healthy and resilient Gulf communities.

3 Advance understanding of the Gulf of Mexico region as a dynamic system with complex, interconnecting human and environmental systems, functions, and processes to inform the protection and restoration of ecosystem services.

Enhance Oil System Safety

GOAL 1 Foster innovative improvements to safety technologies, safety culture, and environmental protection systems associated with offshore oil and gas development.

The oil and gas industry in the Gulf of Mexico region and elsewhere in the United States is a complex, trillion-dollar industry operating in a changing technological and regulatory environment. Major offshore production in the Gulf of Mexico has been shifting into ultra-deep waters (deeper than 5,000 feet) and more hostile environments. Coupled with technological advancements in locating hydrocarbons and accessing them, these resources create increased production capabilities but greater risks. At the same time, infrastructure supporting production in areas developed earlier, as well as onshore transportation, is aging, which also creates new risks. These risks and their associated consequences became apparent during the *DWH* oil spill and led to significant changes in government oversight and to an increased focus by industry on improving safety.

OBJECTIVES   2015-2020

- Partner with industry, government, and academia to explore key factors to prevent future blowouts, oil spills, and accidents and enhance safety culture.

- Explore models of decision-support systems for safe and environmentally sustainable offshore oil and gas development, disaster response, and remediation options.

- Provide research opportunities that improve understanding of how social, economic, and environmental factors influence community vulnerability, recovery, and resilience.

- Support research, long-term observations and monitoring, and information development to advance understanding of environmental conditions, ecosystem services, and community health and well-being in the Gulf of Mexico.

- Support the development of future professionals and leaders—in science, industry, health, policy, and education—who apply cross-boundary approaches to critical issues that span oil system safety, human health, and environmental resources.

- Identify opportunities for knowledge transfer between the Gulf of Mexico and other U.S. outer continental shelf regions.

- Support activities to improve understanding and use of scientific information by the public and policy makers in decisions related to environmental stewardship, human health improvement, and responsible oil and gas production.

In today's increasingly complex offshore oil and gas exploration and development environment, stakeholders recognize that safety risks must be assessed and managed from a systemic perspective that encompasses the human, organizational, and technological factors that affect safety during exploration, production, and the transportation of hydrocarbons from oil fields to coastal regions. Innovation in science and engineering, and continuing training for operators, drilling contractors, and service providers, will be necessary to manage risks safely in these high-pressure, high-temperature, frontier areas.

The Program's contributions to this effort are likely to involve partnerships with industry, relevant federal agencies, educational institutions, and existing safety-oriented centers and institutes. The Program will emphasize systemic approaches, the prevention of harm and reduction of risk, organizational science behind safety, protection of worker health, and the connectivity and interdependence of the industry, people, and communities.

Other Program activities could address foundational research supporting the evolution of technology for the increasing challenges of new oil production environments; issues at the human–technology interface such as decision-making systems, monitoring, responses to

THE EVOLUTION OF OFFSHORE DRILLING AND CHALLENGES AHEAD

The first freestanding offshore oil platform, built in 1938, stood barely offshore in 14 feet of water, and few at the time could have imagined ever drilling in more than 10,000 feet of water as is done now. But technological change and the value of the resource allowed extraction to move deeper and farther offshore.

What challenges lie ahead? Now that much of the more readily accessible oil has been found, new offshore exploration and production activities increasingly take place in frontier areas such as the Arctic and the ultra-deepwater Gulf of Mexico. The challenges of operating in these harsh environments are significant. In the Arctic, extreme weather conditions, ice, and remote locations impose high project costs and spill response challenges. In the ultra-deepwater Gulf, high-pressure, high-temperature conditions require sophisticated engineering and strong management strategies to coordinate among the many companies involved in drilling a well. Flow rates for an ultra-deepwater well can be extremely high, illustrating both the rewards of producing from these reservoirs and the potential risks if another catastrophic spill were to occur. The region also faces the challenges associated with the aging of the older, closer-to-shore infrastructure.

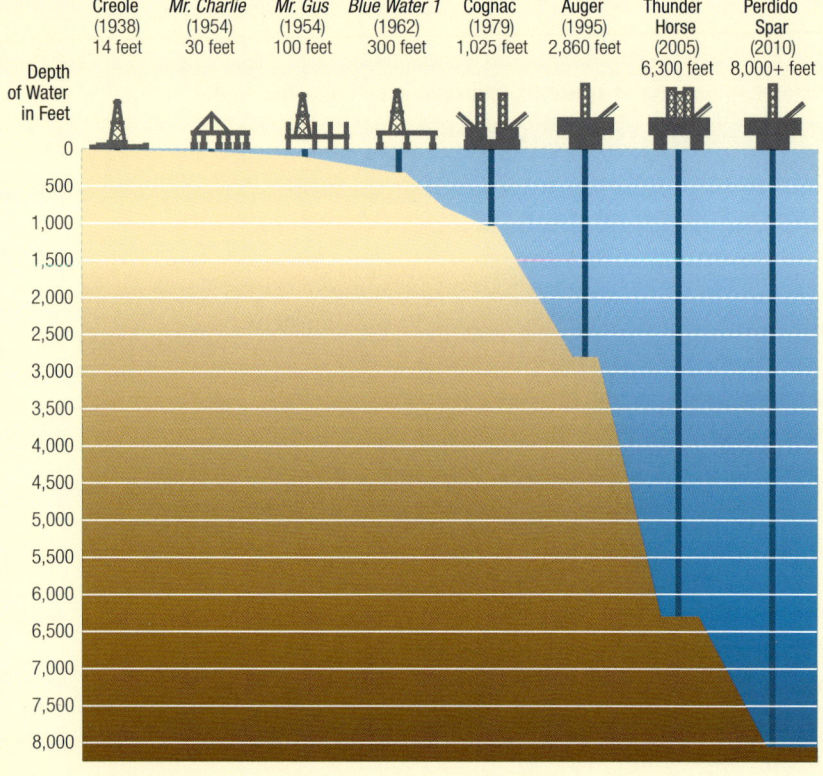

1938—First freestanding offshore platform
(Creole platform, 14 feet of water)
1947—First offshore platform "out of sight of land"
(Kermac 16, 15 feet of water)
1954—First mobile offshore drilling unit
(*Mr. Charlie*, 30 feet of water)
1954—Development of jack-up rigs
(*Mr. Gus*, 100 feet of water)
1956—Development of submersibles
(*Kermac 46*, 80 feet of water)
1962—First semi-submersible drilling vessel
(*Blue Water 1*, 300 feet of water)
1972—First 3-dimensional seismic survey
1979—First true deepwater platform
(Shell's Cognac platform, 1,025 feet of water)
1970s-1980s—Beginning of the deepwater play
(*Discoverer Seven Seas* sets drilling records)
1995—Deepwater becomes truly viable
(Shell's Auger platform, 2,860 feet of water)
2005—Push into deepwater
(BP's Thunder Horse, 6,300 feet of water)
2010—First production from the Lower Tertiary
(Shell's Perdido Spar, 8,000+ feet of water)
2013—World record set for water depth
(Transocean's *Dhirubhai Deepwater KG1*, 10,411 feet of water)

SOURCE: Adapted from Priest (2013).

problems, and critical decisions at critical junctions; and approaches to reduce the impacts of spills and blowouts using improved environmental monitoring of offshore conditions and natural resources around platforms, pipelines, and shipping lanes. Program activities could contribute to progress in training future generations of students, workers, and regulators; facilitating knowledge transfer between academia and industry; and providing a neutral venue for discussions.

Although spill and blowout prevention is a central tenet of oil system safety, the responsibilities associated with offshore production also include preparation for containment, response, and restoration. The Program will seek opportunities to assist in advising industry and government on the best available science and tools. Improving planning for spill response and decision making are also possible areas where the Program's convening functions might assist in bringing together industry, government, and other stakeholders. For example, one area of opportunity that draws interest from multiple public- and private-sector stakeholders is improving understanding of major Gulf currents, such as the Loop Current and its eddies, which can influence the safety and timing of operational activities such as drilling or moving the platforms.

Improve Understanding of Human Health–Environment Connections

GOAL 2 **Improve understanding of the connections between human health and the environment to support the development of healthy and resilient Gulf communities.**

Health is more than lack of illness. Human health, for the Program's purposes, is defined as a "state of complete physical, mental, and social well-being and not merely the absence of disease or infirmity." The environments in which people live, work, and play are connected to health and well-being in a variety of ways, as exemplified by a wealth of ecosystem services in the Gulf region. Coastal wetlands, for example, regulate water quality, reduce the severity of storm impacts, and provide food and recreational opportunities. Other ecosystem services provide highly valued resources such as fish, oil, and natural gas that are extracted by major employers in the region and provide an economic base critical to community and individual well-being. These are just some of the many connections between human communities and their surrounding environments that influence health.

The impacts of the *DWH* oil spill on communities and ecosystems in the Gulf region illustrate the need to better understand these connections. Closure of fishing grounds, for example, threatened the livelihoods of individuals and caused disruptions in entire communities associated with the seafood industry (particularly those also involved in subsistence fishing). More broadly, uncertainty about exposure to and the health and environmental impacts of spill-related contaminants continues to drive public concern about the long-term effects of the spill. Mental and behavioral

health effects are of particular concern, with lessons from the *Exxon Valdez* spill and other disasters suggesting the potential for long-lasting impacts on community recovery. Over the next 30 years, natural and man-made disasters, climate change impacts, and other environmental stressors will present similar, complex challenges to the physical, mental, and social well-being of communities in the Gulf and other continental shelf areas. Understanding the interrelationships among health, ecological, and economic impacts of disasters and other environmental stressors will be crucial to addressing these challenges.

Resilience is a concept used by a variety of disciplines—from engineering to sociology—to describe the capacity of a system to absorb and recover from a disturbance. Resilient human communities anticipate risk, limit impacts, recover quickly, and successfully adapt when faced with adverse events and change. In the Gulf and other coastal regions, resilience is heavily influenced by interactions between human communities and their natural environment, yet these linkages are often not well understood. The Program will seek to foster relevant research across disciplines—including physical, biological, social, and health sciences—to advance understanding of factors that influence the vulnerability, recovery, and resilience of ecosystems and communities. Integrative research to examine these factors—including feedbacks between human systems (health, social, and economic dimensions, among others) and ecosystems—will help communities to better anticipate and respond to disasters and other environmental stressors.

The Program's focus on linkages between the environment and human health will include efforts to improve capacity to detect, assess, and communicate about environmental health risks in ways that support the development of healthy and resilient communities. National, state, and local efforts to improve community resilience have focused on preparedness for, response to, and recovery from all hazards. In addition to improving the understanding of factors influencing community resilience, the Program could contribute to these efforts in a variety of ways, including identifying baseline information needed to track the effects of future disasters and other environmental disturbances; identifying variables and mechanisms that can help address public concerns about health, air and water quality, and the long-term safety of seafood; and improving capacity to prepare for and recover from future adverse events. The Program will seek to provide information that can guide decisions by the public and policy makers and to advance scientific understanding. Additionally, the program will work to support the development of health, scientific, community, and policy leaders who can address complex issues at the intersection between human and ecosystem health.

Advance Understanding of the Gulf of Mexico as a Dynamic System

GOAL 3 **Advance understanding of the Gulf of Mexico region as a dynamic system with complex, interconnecting human and environmental systems, functions, and processes to inform the protection and restoration of ecosystem services.**

The underlying challenge of inadequate ecosystem understanding that was faced during the *DWH* oil spill in 2010 still exists: If there were to be another major blowout or industry-related disaster, is there an adequate base understanding of current conditions to enable adequate response, understanding of damages, and recovery? Is there adequate understanding of systems, functions, and processes and interconnections to inform response, and ultimately recovery? Given the myriad stressors, including the impacts of climate change, is there adequate understanding of key variables to track and anticipate change, and use the information to inform decision making?

Much research has been done on the Gulf of Mexico as a region and ecosystem. However, the scale and complexity of the Gulf leave many unanswered questions, especially about how this large ecosystem functions, how it will change with time, and how it might be affected by perturbations such as those introduced by oil and gas exploration and production as well as severe weather events and changing climate. Scientific research across the physical, biological, social, and

health sciences can provide an objective basis for advancing understanding of the phenomena and processes that define and shape the Gulf ecosystem. It also can help project how the Gulf ecosystem will respond to change—whether slow chronic changes such as rising temperatures or episodic changes from hurricanes or blowouts and oil spills.

Because other parts of the United States also grapple with the interplay of environmental change and energy production, the Program offers an opportunity for research of national significance and relevance and for transfer of knowledge among U.S. regions and other nations addressing similar issues. A critical tool to assist in advancing understanding of the Gulf system will be environmental monitoring. Environmental monitoring, including observations, measurements, analysis, modeling, and interpretation, will require broad, coordinated, and ongoing efforts. To advance the knowledge of the interconnectivity between environmental and human health, similar advances are needed to adequately monitor human health parameters and how coordinated and ongoing efforts can be promoted. The Program will seek opportunities to advance coordinated planning, technological innovation, and integration of data, especially in ways that advance a regional and ecosystem-oriented approach.

Program Objectives

In its initial 5 years of work (2015-2020) the Program will begin to pursue the following program objectives:

- Partner with industry, government, and academia to explore key factors to prevent future blowouts, oil spills, and accidents and enhance safety culture.

- Explore models of decision-support systems for safe and environmentally sustainable offshore oil and gas development, disaster response, and remediation options.

- Provide research opportunities that improve understanding of how social, economic, and environmental factors influence community vulnerability, recovery, and resilience.

- Support research, long-term observations and monitoring, and information development to advance understanding of environmental conditions, ecosystem services, and community health and well-being in the Gulf of Mexico.

- Support the development of future professionals and leaders—in science, industry, health, policy, and education—who apply cross-boundary approaches to critical issues that span oil system safety, human health, and environmental resources.

- Identify opportunities for knowledge transfer between the Gulf of Mexico and other U.S. outer continental shelf regions.

- Support activities to improve understanding and use of scientific information by the public and policy makers in decisions related to environmental stewardship, human health improvement, and responsible oil and gas production.

3
STRATEGIES FOR LASTING BENEFIT

Long-Term, Cross-Boundary Perspective   33

Science to Advance Understanding   34

Science to Serve Community Needs   35

Synthesis and Integration   36

Coordination and Partnerships   38

Leadership and Capacity Building   39

The Program arises out of a seminal event, the *DWH* oil spill, and its many human and ecosystem impacts. Given its $500 million endowment and 30-year duration, the Program presents an extraordinary opportunity to tackle large, complex issues at a regional scale and over the long term. It has the potential to have significant impact and offers a rare opportunity to, over time, address a wide variety of future-oriented challenges.

Taking advantage of this opportunity requires a strategic approach. The planning process identified six overarching strategies that can steer the Program toward producing lasting benefits. These are key opportunities where the mission of the Program aligns with the strengths of the National Academies and where the 30-year duration and long-term perspective hold special potential for cumulative impacts.

Based on community input and Advisory Group deliberations, six key opportunities for the Program to achieve lasting benefit were identified.

Long-Term, Cross-Boundary Perspective

The 30-year duration assigned to the Program offers an unparalleled opportunity to implement future-oriented and long-term initiatives. Thus, as a next step in program development the Program will attempt to select several areas of work that specifically take advantage of the long-term perspective. Similarly, the agreements signal that the Program should not focus on determining the impacts of the *DWH* oil spill, a task being undertaken by many others, including the court system. Rather, it should look toward the future—toward preventing such disasters, minimizing adverse impacts of offshore energy production, and ensuring that the Gulf of Mexico ecosystem and surrounding human communities are resilient to shocks and long-term changes.

Another distinctive feature of the Program is geographic. The mandate given to the Program directs it to focus on the Gulf of Mexico and other U.S. outer continental shelf regions that support oil and gas production. The Program will interpret this scope broadly to mean the Gulf and any U.S. marine and coastal region where human communities, ecosystems, and oil and gas exploration and development interact.

Although the central focus of the Program will remain the quest to contribute to a healthier and more resilient Gulf region, the Program is directed to operate more broadly where similar issues

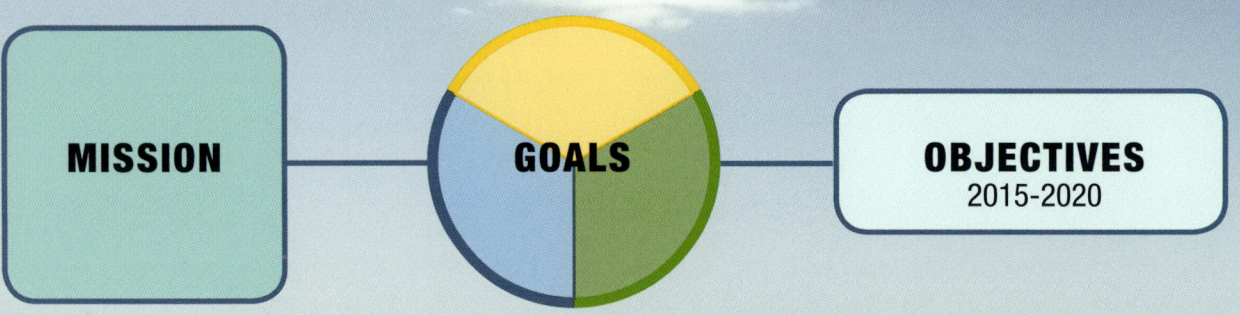

STRATEGIES FOR LASTING BENEFIT

- Long-Term, Cross-Boundary Perspective
- Science to Advance Understanding
- Science to Serve Community Needs
- Synthesis and Integration
- Coordination and Partnerships
- Leadership and Capacity Building

are faced in other outer continental shelf areas with existing or potential energy development. It will tap relevant expertise and insights from outside the Gulf and approach issues holistically. Other states (such as Alaska) and other nations face decisions about proceeding with offshore energy development and could benefit from Gulf region activities. Other areas in the United States and around the world have experience in offshore energy development and oil spills that could provide important information. Thus, work that transfers knowledge to or from other places in the United States, the Gulf of Mexico, or other nations is possible within the scope of the Program. Collaborative work that demonstrates such linkages will be valued. Over time, the Program expects to include international perspectives and collaborations.[5]

Finally, the three areas of responsibility (oil system safety, human health, and environment) assigned to the Program are extremely broad. In seeking focus, the Advisory Group determined that the Program's most valuable contributions are likely to come at the intersections of these realms. Untapped opportunities exist at these intersections to bring disparate perspectives, disciplines, and sectors together and to spur innovation.

Science to Advance Understanding

A fundamental purpose of the Program is to bring the best expertise in science, engineering, technology, and health to help advance understanding of the Gulf of Mexico and the use of that information in decision making. Important research questions are varied, dynamic, and

---

[5]Future documents and the Program's website (www.nas.edu/gulf) will provide details on who is eligible to apply under announced funding opportunities. For the program's first 2 years, awards will be made only to U.S. institutions, citizens, and permanent residents; funded institutions may collaborate with individuals outside the United States.

interrelated—from understanding deep-ocean circulation, to developing new approaches to avoid future blowouts or oil spills, to understanding how communities could become more resilient to disasters and future change. The Program will encourage cross-boundary work across disciplines, across geographic borders, and across perspectives.

The Program will foster research that explores the links among people, ecosystems, and energy development, particularly the ways in which humans affect the environment and how the environment affects people (especially health and well-being). It will seek to create opportunities that encourage new ways of thinking and potentially transformative science and technology.

Science to Serve Community Needs

The Program will seek to foster science that serves the needs of the region's numerous and diverse communities. This science will include "translational research"[6] or "actionable" science that is focused on the ways in which new knowledge can be used by the public, resource managers, program managers, community planners, and other decision makers and how the public can

---

[6]Translational research is research that strives to translate scientific discoveries into tangible human benefit. It is practiced in many fields, but is perhaps best known in medicine in a movement to move scientific findings "from bench to bedside" (Sung et al., 2003).

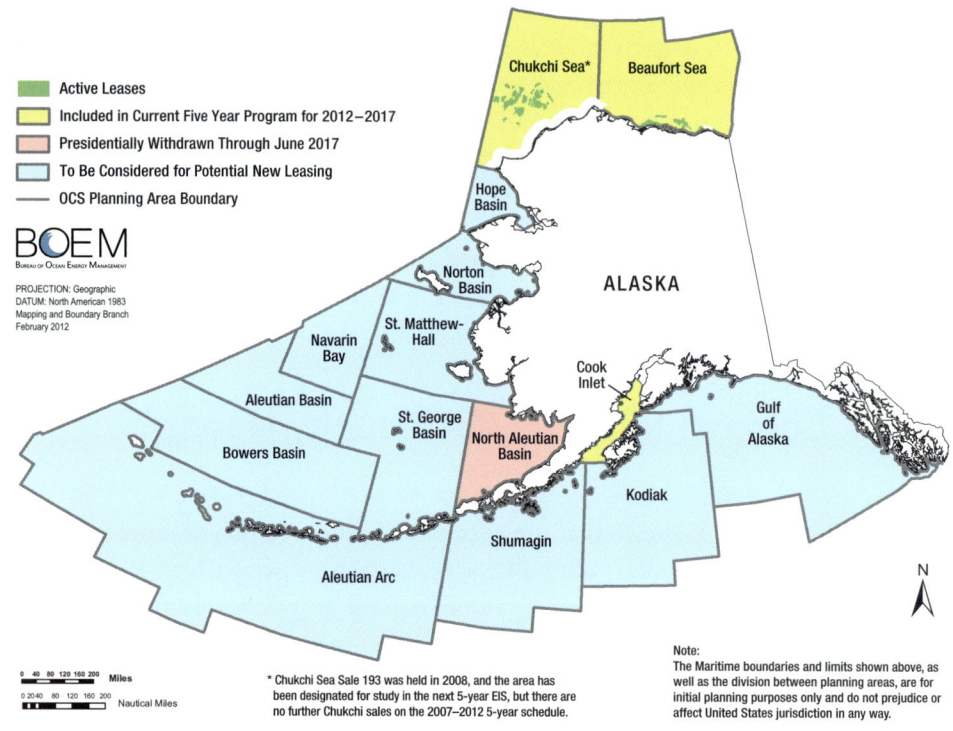

inform priorities for new science initiatives. This approach requires an understanding of the concerns and priorities of communities and policy decision makers and the means to transfer knowledge from basic research settings to those associated with the implementation of that knowledge. For example, research to identify opportunities to speed postdisaster recovery can be coupled with research on policy implementation that would encourage decision makers to take advantage of these opportunities.

Translational research includes efforts to develop and encourage the adoption of best practices to maximize the impact of research. A particularly promising option for the Program in this area is the identification of best scientific practices, such as those for design, monitoring, and evaluation of restoration projects. Generally, the Program will emphasize regional and ecosystem approaches over state or locally focused activities, striving for insights with broad applicability.

Synthesis and Integration

The Program recognizes that there is a great need for increased attention to the synthesis and integration of data and information, especially across disciplines. Scientific synthesis is the

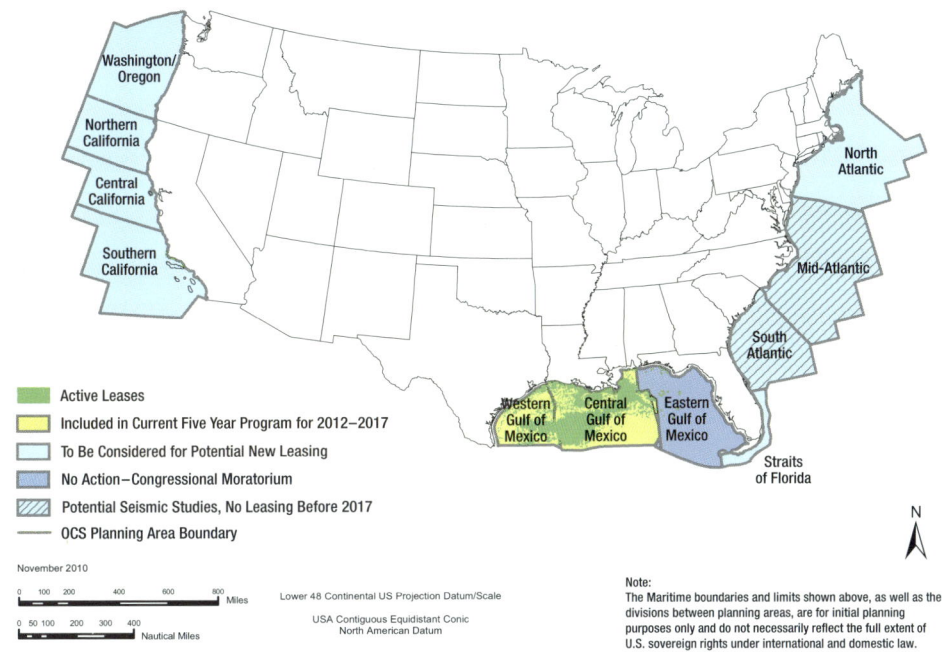

The U.S. Department of the Interior, through the Bureau of Ocean Energy Management (BOEM), oversees an Outer Continental Shelf (OCS) Oil and Gas Leasing Program that establishes the areas that may be leased for offshore energy development. The most recent leasing window covers a 5-year period (2012–2017). Currently, 218.94 million acres of the U.S. OCS are open for development: (A) 125.19 million acres in Alaska and (B) 93.75 million acres in the Gulf of Mexico. Seismic studies are allowed in the mid- and south Atlantic, but no leases will be issued in this area before 2017. SOURCE: BOEM.

and applicability of scientific research. Modes of scientific synthesis include integration of methods, data aggregation, reuse of results, and conceptual synthesis (Sidlauskas et al., 2009).[7] There may be opportunities to explore "big data" to gain new insights. Traditional consensus studies and workshops conducted under the auspices of the National Academies often are designed to synthesize the best available information and translate new understanding into action.

Different modes of synthesis can be applied within or across disciplines and professional sectors to improve understanding of the Gulf of Mexico as an ecosystem. For example, aggregated archival oil industry and Department of the Interior data collected for leasing and exploration activities during the 1970s and 1980s could be compared to current data collections to develop environmental reference points in assessing disaster impacts and long-term changes. Assessing health effects from environmental contamination as a result of oil and gas exploration might be done by linking datasets across disciplines.

The Program will seek opportunities to facilitate the process through leadership, policies, and investments. For example, the Program may invest in research supporting data systems of the

---

[7] In data aggregation, multiple datasets are merged and analyzed, typically to address questions at new and larger scales. The reuse of results involves using data in a new context, as in the case of meta-analyses. With integration of methods, two or more methods are combined to create a new analytical pathway. Conceptual synthesis bridges the theories and paradigms that underpin previous studies.

future; help establish the meta-analysis agenda; catalyze partnerships around data generation, management, and use; and assist in supporting the development of technologies, protocols, and standards.

Coordination and Partnerships

The Program recognizes that it is but one program among many operating in the Gulf region. Nevertheless, the status of the NAS as an independent, nonprofit organization with a long history of consensus building allows the organization to serve as a neutral convener of diverse perspectives. In this role, the Program can provide leadership and participate in efforts that facilitate coordination and build partnerships, including on emerging or controversial topics. For example, the Program intends to be a catalyst in the area of environmental monitoring, where the needs are expansive and expensive and thus no one organization can succeed alone.

The Program will allocate resources to conduct targeted workshops and assessments that could facilitate partnerships. Program staff will continue to interact regularly with other related Gulf region funding organizations to ensure communication and seek opportunities for coordination and collaboration. The Program recognizes the importance of engaging with a diverse range of stakeholders so its work will be informed by those who live and work in the Gulf and understand the region's needs and challenges. Over time, the Program will continue to improve communications and engagement methods so that it contributes to and benefits from a two-way flow of information. The Program has a special opportunity to look at the Gulf region holistically, as a region and not as separate states, and bring a national dimension into the dialog.

Leadership and Capacity Building

Through a range of activities over the 30 years, the Program will seek to invest in capacity building—providing opportunities for emerging academic and community leaders, state and regional decision makers, students, and institutions to develop skills, competencies, and capabilities that are needed to solve problems, spark innovation, and establish more sustainable systems, economies, and communities. By providing opportunities for leadership development, a next generation of individuals and institutions can emerge prepared to think broadly and innovatively to resolve problems in complex, multistressed social–ecological systems and enhance community resilience.

The Program recognizes the diversity of people and cultures in areas such as the Gulf region and other U.S. outer continental shelf areas with existing or potential oil and gas development. It will seek ways to engage and facilitate the participation of institutions, organizations, and diverse populations in relevant work. In addition, workforce demands are likely to evolve significantly over the next 30 years, increasing the need for a workforce that is flexible and can adapt to changes in technology, safety requirements, and workplace demands.

4
INITIAL AND FUTURE ACTIVITIES

Initial 2015–2016 Activities   41

Next Steps for 2015–2020   45

Next Steps in Identifying Long-Term Initiatives   46

With its mission, goals, and objectives articulated, the Program can begin moving from vision toward implementation. Under the guidance of the Advisory Group, the Program has developed a suite of initial activities, with details to be announced in fall 2014 and first funds awarded in 2015. These initial activities are expected to evolve and other, to-be-determined activities will be added over the first 5 years. Responsibility for continued development of larger- and longer-scale activities, oversight, evaluation, communications, and other Program functions will be the role of a new Advisory Board, to take charge in fall 2014.

The initial activities are primarily short term and foundational, to allow activity to begin while planning for larger and longer-term activities continues. The Program's 30-year time frame and the unique role played by the National Academies hold special potential. What can this Program cause to happen that would not have happened otherwise? Can the Program offer opportunities that inspire innovative, integrated thinking about how to address the challenges faced by the people along the Gulf Coast?

Given that the Program's funds are to be expended over 30 years, which will constrain annual expenditures, relevance to the Program's mission and announced themes and the potential to add value will be important criteria when judging potential activities, in addition to scientific excellence and technical merit. Program areas of emphasis, research themes, and specific activities will evolve over time. In general, the Program is not expected to supplement existing, separately sponsored work. Nor will the Program support on-the-ground restoration activities, which are the purview of other significant programs. Proposals must address designated themes during designated cycles.

Initial 2015–2016 Activities

In fall 2014, the Program will issue its first calls for applications for three initial, short-term funding opportunities: exploratory grants, research fellowships, and science policy fellowships. These opportunities will be funded in 2015 and again in 2016. An opportunity to foster integration and synthesis of monitoring data will open in early 2015. To foster coordination and the sharing of current knowledge—for example, through conferences, workshops, and collaboration meetings—the Program may make available a small number of collaboration/coordination grants in 2015. In addition, expert consensus studies of value to the Gulf region, planning meetings to inform the Program's future activities, and workshops and other mission-relevant activities are expected to be developed, engaging a wide range of stakeholders from the Gulf and elsewhere.

Exploratory Grants

The initial suite of exploratory grants in 2015 and 2016 are intended to catalyze innovative thinking around selected issues identified during planning that support the Program's goals and objectives. The exploratory grants are designed to provide seed money for research in its early conceptual phase or for activities that can accelerate the development of novel approaches or the transition from concepts to testing. This funding opportunity will allow innovators to test ideas, collect preliminary data, or change direction as a result of insights from exploratory projects. The grants also could support the use of novel approaches, application of new expertise, or engagement of nontraditional or interdisciplinary perspectives to break new ground on an old or new problem.

In fall 2014, the exploratory grant competition will seek requests for applications on two topics. The awards, to be made in 2015, will total about $2 million, with 15 to 20 awards expected at an average award size of about $100,000. The Program will welcome proposals from individuals or teams of investigators from eligible U.S. institutions. The 2015 topics will be:

Exploring Approaches for Effective Education and Training of Workers in the Offshore Oil and Gas Industry and Health Professions. The nation's middle-skilled workforce includes workers in occupations that require considerable skill but not advanced degrees. In the Gulf of Mexico and outer continental shelf, these workers in the offshore oil and gas industry and health professions play key roles in maintaining the safety of people and the environment and in improving disaster preparedness and response. The focus of this opportunity is to explore new and innovative approaches for educating and training middle-skilled workers in the oil and gas industry and health professions (including individuals working as emergency responders and environmental specialists). Such approaches would specifically leverage the growing evidence base about how people learn to improve safety in job functions and operations and to improve disaster preparedness and response.

Linking Ecosystem Services Related to and Influenced by Oil and Gas Production to Human Health and Well-Being. The Gulf of Mexico delivers a broad suite of ecosystem services, including the provision of seafood, stabilization of coastal habitat, and recreational opportunities. The region also produces oil and gas, which benefit individuals with occupations as well as communities with energy, economic growth, and stability. Managing a diverse portfolio of ecosystem services to meet human needs is a central challenge because pursuing the benefits from one ecosystem service may result in diminishing the quantity or quality of other ecosystem services in the same region. An improved understanding of ecosystem services in relation to the production of oil and gas, their provision under dynamic conditions, and their interconnectedness to human communities would help optimize the multiple services provided by a system, manage trade-offs, and inform decisions in ecosystem restoration. The focus of this opportunity is to advance knowledge in ecosystem services related to or influenced by offshore and coastal energy production and their linkages to human well-being. The funded activities could expand and accelerate the application of ecosystem services to the management and restoration of the Gulf of Mexico and other ecosystems in the outer continental shelf.

An exploratory grant competition also will be held in fall 2015, for funding in 2016. The extent of funding in award year 2016 will be determined on the basis of program budget and the number and quality of applications received in fall 2014. The topics in award year 2016 are expected to include

- Innovative approaches to developing scenario planning and decision-support systems to cope with crises;
- Connecting data about environmental conditions with individual and population health data to foster transdisciplinary research; and
- Building resilience in human and environmental systems of the Gulf of Mexico and other offshore energy-producing regions.

Fellowships

The Program is committed to the long-term task of capacity building in the Gulf, including the development of future generations of scientists, engineers, and health professionals prepared to work at the intersections of oil system safety, environmental resources, and human health in the Gulf region and to think holistically and at the community and ecosystem levels about the region's challenges.

As a first step toward enhancing the breadth and leadership capacity of leading early-career science, engineering, and health professionals, the Program will initiate two new fellowship opportunities and participate in one ongoing NAS fellowship program. The fellowship programs will encompass participants from a broad range of disciplines, including the social and behavioral sciences, health and medicine, engineering, the earth and life sciences, and relevant interdisciplinary fields. The fellowships will be awarded to applicants whose research or work relates to the mission and objectives of the Program. Mentoring will be a critical component of both new fellowship programs. The three types of fellowships are:

Early-Career Research Fellowships will recognize professionals at the critical pretenure phase of their careers for exceptional leadership, past performance, and potential for future contributions to improving oil system safety, the environment, or human health. These 2-year fellowships will be awarded to tenure-track (pretenure) faculty (or the equivalent) at colleges, universities, and research institutions. Fellowship funds will be used primarily for research-related expenses and professional development.

Science Policy Fellowships will contribute to leadership development and capacity building by providing recipients with a valuable educational experience at the science–policy interface. Fellows will spend 1 year on the staff of a state legislature; state environmental, natural resources, oil and gas, or public health agency; or regional offices of relevant federal agencies. Fellows will participate in and contribute to the state or federal policy making process. Depending on placement, duties are likely to include developing and analyzing policy, writing policy memos, and drafting legislation. The fellowships will be awarded to graduate and professional school students and those who have completed their graduate studies (M.A./M.S., Ph.D., Sc.D., M.P.H., M.D., D.V.M.) no more than 5 years before beginning the fellowship. Fellowship benefits will include a monthly stipend and support for professional development courses (e.g., science communication or project management) and professional travel.

The Christine Mirzayan Science & Technology Policy Graduate Fellowship Program.
This existing National Academies fellowship program brings two dozen talented early-career professionals from a variety of disciplines to Washington, DC, each year for 12 weeks and provides a broad introduction to the role of science in the federal government.[8] The fellow hosted by the Program will work directly with staff and Advisory Board members and have broad opportunities to learn about the role of science at the federal level and contribute directly to the Program's work in the Gulf of Mexico.

Integration and Synthesis of Monitoring Data

In 2015, the Program anticipates announcing its first competition in the area of environmental monitoring. The opportunity will seek to demonstrate the untapped potential of existing long-term environmental observations and monitoring data. Applicants will be challenged to propose hypothesis-driven projects that identify and synthesize existing data related to either the deep Gulf or ecosystem services for restoration and management themes. Funded projects are expected to generate outputs that can lead toward better-informed decision making, translation into human benefits, or other actionable outcomes. Grants will be made to define data needed to address the research question; identify existing data, their location, and access to the data;

---

[8]For information about the Mirzayan Fellowship Program, including purpose, application procedures, and deadlines, please see http://sites.nationalacademies.org/pga/policyfellows.

identify parties necessary for collaboration; and demonstrate the feasibility of the project. Special consideration will be given to projects that integrate data from different disciplines. If a funded project shows exceptional promise, some awardees may receive additional (phase two) funding to process and integrate the data and potentially expand the project.

Next Steps for 2015–2020

The Program will start small as it receives funds in its first few years, yet it will evolve relatively quickly to include a range of activities. The development of these activities will be informed by workshops, community interactions, and the guidance of the Advisory Board. Although the fellowships might evolve as the Program learns from the first year of experience, they are expected to be offered each year for at least 5 years until the first program evaluation in 2020. A cycle of regular research funding opportunities will be developed, with changes in focus and scale.

Program initiatives are expected to be informed by ongoing work. For example, the Program's oil system safety portfolio will be informed by a recently initiated study on improving safety culture by the Transportation Research Board of the National Academies. Future directions will also be informed by three 2014 opportunity analysis workshops that will generate a range of ideas for the Program to consider. Topics under discussion at these workshops include the following:

Education and Training:
- Education and training pathways for the Gulf region's middle-skilled workforce in the hydrocarbon and environmental restoration industries and the health professions; and
- Opportunities to develop capacity to meet future middle-skilled workforce needs in the Gulf region.

Environmental Monitoring:
- Use of observations and monitoring to better understand the deep ocean and connectivity to the coast; and
- Use of observations and monitoring to support ecosystem services modeling to inform restoration activities.

Community Resilience and Health:
- Strengthening community resilience, health, and well-being; and
- Improving detection, assessment, and communication about environmental health threats.

The 2014 workshops are intended to be in-depth discussions where participants can explore what activities are currently being done, what plans exist, and which specific opportunities should be pursued by the Program. The workshops are expected to lead to the development of additional

Program activities and opportunities for the research community. Interaction between Gulf-focused experts and those working on other U.S. outer continental shelf regions is a possible topic for a future workshop.

During 2015–2020, other research themes and initiatives will be designed and implemented, including activities such as support for larger, consortia-driven research projects; development of "method tool boxes" that allow researchers, especially health researchers, to quickly establish reference measurements after future events; student competitions; or implementation of educational programs. Over time, the areas of focus are expected to evolve with changing priorities.

Use of Traditional National Academies' Modes of Operation

During 2015–2020, the Program will begin to take advantage of some of the traditional modes of operation for the National Academies as a neutral convening body to bring together top experts in science, engineering, and health. This can take the form of consensus studies with recommendations, workshops designed to explore topics or gather divergent views, and other activities that bring together targeted groups of professionals for in-depth discussion. The Program intends to take advantage of this core institutional strength and, over time, will focus the traditional committee and workshop functions on issues of wide interest to the Gulf and other offshore energy-producing regions. For example, given the significant restoration efforts under way and being planned for the Gulf region, the Program anticipates a project to identify best practices for monitoring and evaluating restoration projects.

Next Steps in Identifying Long-Term Initiatives

The Program recognizes the need to demonstrate its intent by announcing initial funding opportunities, but the Program's greatest impact likely will come from larger and more far-reaching activities that will take longer to plan and implement. During its meetings to gather input, the Advisory Group heard suggestions of many large and longer-term initiatives that the Program could pursue, from specific activities to visionary undertakings. One of the first challenges for the new Advisory Board will be to explore these possible activities and engage in the necessary, detailed planning.

As it develops, the Program hopes to inspire transformative changes that can improve oil system safety, human health, and the environment in the Gulf. Identifying and developing larger, more far-reaching ideas will require further exploration and planning. One example is the idea of supporting the design and development of an interdisciplinary or cross-boundary center to specifically address oil system safety, human health, and environmental resources in a holistic manner. Many models of interdisciplinary centers and centers of excellence exist, some of which have a defined life span while others are established to last indefinitely. Similarly, the area of environmental monitoring is ideal for the Program's 30-year duration, in particular, the opportunities for integration and synthesis of data. Another example is what role the Program might play in advancing the use of "big data," especially from environmental monitoring, to tease out useful information.

To encourage cross-boundary innovation and collaboration, the Program will continue to conduct additional planning, program development, and community engagement. The Ideas Lab administered by the National Science Foundation or the Keck Futures Initiative administered by the National Academies are examples of opportunities that bring together experts from a range of disciplines to develop bold and innovative approaches to address critical societal challenges. Similarly, a number of centers in the United States support synthesis research that brings together interdisciplinary teams of investigators to distill existing data, methodologies, theories, and ideas from different sources and research fields to increase the applicability of research results, address issues on a broad scale, and generate new knowledge.

Any large investment requires careful consideration of purposes, conceptualization, planning, implementation, and sustainability over the long term. In 2015, the Program will identify a few, select grand challenge ideas and will hold exploratory discussions that lead to planning this next level of activity. The incoming Advisory Board will work to identify future potential activities that align with the program's mission and objectives, play to the strengths of the National Academies, and increase the Program's impact.

5
PROGRAM OPERATIONS

Scientific Integrity 49

Coordination and Communication 50

Planning and Evaluation 51

Being housed within the NAS gives the Program advantages that come with being placed in a long-established institution with significant resources, established procedures and protocols, and accumulated and transferable knowledge. It allows the Program access to a great diversity of volunteers and staff expertise, as well as to existing infrastructure and support systems. The Program will take advantage of existing institutional support and expertise whenever possible, from existing financial systems to oversee and manage funds, to the National Academies Press for publication needs, to the Office of Government and Congressional Affairs as an interface with Congress and the Executive Branch of the federal government.

Scientific Integrity

A fundamental purpose of the Program is to facilitate the advancement of knowledge and the application of science to address Gulf region challenges. All activities of the Program will be conducted to meet the highest standards of scientific integrity. All staff, volunteer advisers, fellows, and grantees have a responsibility to use the funds wisely. To ensure scientific integrity in preaward processes, the Program will develop peer review procedures modeled generally on National Science Board (2005) evaluation protocols. To avoid conflicts of interest in the selection process, independent reviews will be performed by scientific peers not affiliated with institutions who propose projects. The Program peer review process and criteria for selection will be detailed in each request for applications and on the Program's website.

To continue the emphasis on scientific integrity throughout the award period, the Program will ask all researchers, trainees, and fellows to comply with professional standards as defined by the NAS report *On Being A Scientist: A Guide to Responsible Conduct in Research* (IOM et al., 2009). Dissemination of results through professional and public channels will be strongly encouraged. To promote data verification and reproducibility, and to maximize the value of the data generated by encouraging syntheses and reuse of data, data generated by the Program's grantees should be made available in a timely manner and to the greatest extent possible, subject to any institutional review board or legal restrictions. Procedures will be described on the Program website and in awardee agreements. The Program will abide by the NAS's institutional conflict-of-interest policies and procedures for board members, other volunteers, and staff.

Coordination and Communication

The Program will maintain relationships with a range of federal, regional, state, nonprofit, and industry leaders and organizations to ensure that Program activities are widely known, that opportunities for leveraging impact are identified, and that, cumulatively, Program efforts are as coordinated as is feasible given the diversity of mandates. Opportunities for two-way dialog will be essential, as will engagement and transfer of knowledge among those working on similar issues in other outer continental shelf regions. Given the Program's mission, building relationships with industry and safety-focused centers and consortia will be essential. Mechanisms such as conferences, workshops, outreach meetings, or Web-based opportunities will be used to communicate Program opportunities and allow input into Program planning over time.

As Program activities develop, communications activities also will advance. The Program will develop a communications plan that identifies how best to communicate both its processes and the products of the work funded, and will work with others who specialize in communications to engage with and serve educators and students, program managers, decision makers, and industries and citizens who will benefit from the results of the Program's work. Successful communications require an understanding of and long-term interaction with a variety of audiences, including

- Science, engineering, and health professionals;
- Managers of related science, engineering, and health programs involved in other Gulf of Mexico work (federal, regional, state, local, industry, and nonprofit);
- Gulf State regional and state-level decision makers with health, education, and environmental or resource management responsibilities;
- The offshore oil and gas industry, including major operators, drilling contractors and service providers, industry working groups, and organizations that address safety issues;
- Federal agencies and programs with research and restoration responsibilities in the Gulf region and in other outer continental shelf areas with relevance to oil and gas production;
- Regional planning and resource management bodies and local governments, community leaders, and key Gulf State decision makers;
- Organizations working at the boundary between producers and users of scientific results;
- Relevant nongovernmental organizations and community leaders; and
- Relevant units of the National Academies devoted to energy, health, or environmental topics of interest in the Gulf region.

Planning and Evaluation

Thirty years is a significant duration for a program, especially in science. The Program is committed to using the principles of adaptive management to guide the Program's operations and areas of focus. The Program's Advisory Board will play a critical role in planning, oversight, evaluation, and strategic course corrections. It will guide the development of Program procedures and systems that are efficient and effective, and that are flexible so that the Program can change to address new developments in science, technology, and priorities. The Program will strive to be transparent in its evaluation processes and will continue to engage with other programs and the region's experts and decision makers in program planning. The Program will evolve as it matures, but several principles will ensure that development of the Program remains true to the vision of the initial Advisory Group. These include emphasizing a future-oriented perspective; encouraging excellence in science; engaging stakeholders in the Gulf region and beyond; catalyzing the development of potentially transformative science and technologies; and encouraging innovation, collaboration, and education.

Role of the Advisory Board

A program of this breadth and financial impact requires thoughtful leadership. Intellectual oversight and program strategic guidance will come from an appointed Advisory Board whose volunteer members will be selected through normal Academies processes to bring expertise and independent judgment to discussions. The Advisory Board will have approximately 18 members who will be appointed to rotating, 3-year terms (with a second term possible). Advisory Board members will bring expertise reflecting the Program's mission and goals; they will include but not be limited to residents of the Gulf States. Participation by members of the NAS, the NAE, and the IOM will continue.

Advisory Board members will meet two to three times per year, either in full or in subgroups, and will work as needed between meetings to set strategic directions for the program and oversee implementation. Advisory Board members will be responsible for

- Providing strategic leadership and identifying emerging issues and novel activities;
- Guiding the development of priorities, themes, and activities and related calls for applications and other opportunities;
- Participating in ongoing Program planning and portfolio balance discussions;
- Serving on working groups and/or subcommittees (e.g., selection of fellows);
- Maintaining relationships with relevant stakeholders, including federal agency staff, Gulf State officials, industry, nongovernmental organizations, and citizen groups;
- Advising on selection of external reviewers;
- Contributing to the award selection process; and
- Contributing to regular Program assessments and reviews, including an annual report and 5-year reviews.

Advisory Board members will be subject to normal NAS conflict-of-interest policies. In certain circumstances, members may be expected to recuse themselves from discussions in the event of a potential conflict. Because of conflict-of-interest rules and the impact they would have on Board composition, Advisory Board members will have input into, but not make, funding decisions. Peer review and Advisory Board review will be used to evaluate and rank proposals in competitive processes; final funding decisions will be made by the NAS. As part of the required 5-year program review, the Advisory Board's structure and function will be evaluated and adjusted as necessary to improve effectiveness.

Program Evaluation and Metrics

Evaluation will be a critical component of the Program because it contributes to promoting self-understanding, accountability, and transparency; it also allows the identification and leveraging of Program strengths. Regular evaluation, including a required 5-year review, will help identify ways to improve operations and enhance effectiveness. Finally, evaluation will allow assessment of progress toward the Program's goals and guide the Program as it adapts to changing needs and priorities.

Evaluation will take place at different time intervals and at the grantee, activity, and Program levels. The agreements that guide the Program require at least yearly reports from grantees on project accomplishments, expenditures, and final outcomes. Required reporting at the grant level will help the Program's staff monitor the grantees' progress, ensure financial accountability, and track deliverables.

At the activity level (e.g., exploratory grants and fellowships), metrics for evaluating success will be defined at the outset of each activity. Having a set of defined metrics will help identify what information needs to be collected, how and when the information will be collected, and how the collected information will be analyzed. Activity operations also will be assessed at the end of each cycle, so that inefficiencies can be identified and procedures modified to enhance operations of the next cycle.

In addition, information on an activity's inputs, outputs, and outcomes will be collected, and all of the grantees' reports for an activity will form the basis of the activity-level evaluation. Using exploratory grants and fellowships as an illustration, examples of inputs, outputs, and expected outcomes[9] are shown in the table below.

TABLE **Illustration of Inputs, Potential Outputs, and Outcomes of Program Activities**

Activities	Inputs	Potential Outputs	Examples of Potential Outcomes and Their Metrics
Fellowships	Amount invested, expertise and assets of institutions hosting the fellows	Numbers of fellows trained	Potential outcome: workforce development Metrics: number of past fellows with policy or research careers in areas relevant to the Program's mission
Exploratory grants for research and development	Amount invested, number of principal investigators (PIs) involved in each grant, expertise of PIs, and assets of PIs' home institutions	Data, research publications (for basic research), patents, or products (for applied research)	Potential outcomes: transformative insight or new knowledge that leads to the development of new products or new practices (basic research), wide adoption of new products or practices (applied research), and wider economic impacts

For many activities, outputs and outcomes might need to be tracked long after projects or fellowships end to gain a complete understanding of their impact. In academia, articles depicting research results (an output) may be published after the termination of activity. An applied-research project might result in a product developed as an output, but the project outcome might be insignificant if the product does not penetrate the market and is not used. Alternatively, market uptake could be slow initially, but the product could become widely used later and have

---

[9] Inputs are the resources needed to make the activity happen. Outputs are the direct result of the activity. An outcome is a change that has occurred as a result of the activity output over the longer term.

a significant impact. Other outcomes might be difficult to establish. For example, linking cause and effect between basic research and its contribution to advancing knowledge that ultimately leads to a wider economic impact could be difficult because of the multiple steps involved in the application of new knowledge.

All activity-level evaluations will contribute critical information to the Program's annual report, which will be a summary of each year's activities, including expenditures and accomplishments. The annual reports will help the Advisory Board determine whether the mix of activities in the Program's portfolio is appropriate and recommend changes such as introducing, terminating, expanding, or reducing activities. The annual report will be produced in February of each year to allow accurate year-end financial summaries.

The activity-level evaluations and the Program-level annual reports will serve as inputs for a 5-year program review to be conducted by a team of outside experts who will assess the Program's progress toward Program goals and objectives. In addition to quantitative data, the evaluation team may consider engaging stakeholders to ensure that the Program's activities and the Program as a whole have proven relevant to their needs.

Given the early stage of Program formation and the complex issues that the Program aims to address, evaluation strategies will seek to provide feedback for continuous learning and program adaptation. Based on the stakeholders' input and the data collected by the Program, the evaluation team will address the following:

- How did the Program perform against its stated goals and objectives?
- Has the Program been responsive to changing needs?
- Did the Program's activities complement and/or leverage other existing activities?
- Does the Program need to alter its strategy/vision to respond to changes in needs or priorities?
- What can the Program do to improve its implementation and impact in the next 5 years?

REFERENCES

Environmental Law Institute and Tulane Institute on Water Resources Law and Policy. 2014. *Deepwater Horizon Restoration & Recovery Funds: How Much, Going Where, For What?* White Paper. Available at http://eli-ocean.org/gulf/files/Funding-DH-Restoration-Recovery.pdf (accessed September 2, 2014).

GoMRI (Gulf of Mexico Research Initiative). 2014. *Gulf of Mexico Oil Spill & Ecosystem Science Conference Report.* Available at http://2014.gulfofmexicoconference.org/wp-content/uploads/2014_GulfConferenceReport.pdf (accessed September 2, 2014).

Ingram, K. T., K. Dow, L. Carter, and J. Anderson. 2013. *Climate of the Southeast United States: Variability, Change, Impacts, and Vulnerability.* Washington, DC: Island Press.

IOM (Institute of Medicine). 2010. *Assessing the Effects of the Gulf of Mexico Oil Spill on Human Health: A Summary of the June 2010 Workshop.* Washington, DC: The National Academies Press.

IOM, NAS, and NAE (Institute of Medicine, National Academy of Sciences, and National Academy of Engineering). 2009. *On Being a Scientist: A Guide to Responsible Conduct in Research,* 3rd Ed. Washington, DC: The National Academies Press.

IPCC (Intergovernmental Panel on Climate Change). 2013. *Climate Change 2013: The Physical Science Basis. Contribution of Working Group I to the Fifth Assessment Report of the Intergovernmental Panel on Climate Change,* T. F. Stocker, D. Qin, G.-K. Plattner, M. Tignor, S. K. Allen, J. Boschung, A. Nauels, Y. Xia, V. Bex, and P. M. Midgley, eds. Cambridge, UK, and New York: Cambridge University Press.

IPCC. 2014a. *Climate Change 2014: Impacts, Adaptation, and Vulnerability. Part A: Global and Sectoral Aspects. Contribution of Working Group II to the Fifth Assessment Report of the Intergovernmental Panel on Climate Change,* C. B. Field, V. R. Barros, D. J. Dokken, K. J. Mach, M. D. Mastrandrea, T. E. Bilir, M. Chatterjee, K. L. Ebi, Y. O. Estrada, R. C. Genova, B. Girma, E. S. Kissel, A. N. Levy, S. MacCracken, P. R. Mastrandrea, and L. L. White, eds. Cambridge, UK, and New York: Cambridge University Press.

IPCC. 2014b. *Climate Change 2014: Impacts, Adaptation, and Vulnerability. Part B: Regional Aspects. Contribution of Working Group II to the Fifth Assessment Report of the Intergovernmental Panel on Climate Change,* V. R. Barros, C. B. Field, D. J. Dokken, M. D. Mastrandrea, K. J. Mach, T. E. Bilir, M. Chatterjee, K. L. Ebi, Y. O. Estrada, R. C. Genova, B. Girma, E. S. Kissel, A. N. Levy, S. MacCracken, P. R. Mastrandrea, and L. L. White, eds. Cambridge, UK, and New York: Cambridge University Press.

IPCC. 2014c. *Climate Change 2014: Mitigation of Climate Change. Contribution of Working Group III to the Fifth Assessment Report of the Intergovernmental Panel on Climate Change,* O. Edenhofer, R. Pichs-Madruga, Y. Sokona, E. Farahani, S. Kadner, K. Seyboth, A. Adler, I. Baum, S. Brunner, P. Eickemeier, B. Kriemann, J. Savolainen, S. Schlömer, C. von Stechow, T. Zwickel, and J. C. Minx, eds. Cambridge, UK, and New York: Cambridge University Press.

LSU AgCenter (Louisiana State University AgCenter). 2012. *Racial and Ethnic Groups in the Gulf of Mexico Region: A Series.* Baton Rouge: Author. Available at http://www.lsuagcenter.com/en/communications/publications/Publications+Catalog/research/Community/RacialandEthnicGroupsintheGulfofMexicoRegionSeries.htm (accessed September 2, 2014).

Mabus, R. 2010. *America's Gulf Coast: A Long Term Recovery Plan After the Deepwater Horizon Oil Spill.* September. Available at http://www.restorethegulf.gov/sites/default/files/documents/pdf/gulf-recovery-sep-2010.pdf (accessed September 2, 2014).

McNutt, M. K., S. Chu, J. Lubchenco, T. Hunter, G. Dreyfus, S. A. Murawski, and D. M. Kennedy. 2012. Applications of science and engineering to quantify and control the *Deepwater Horizon* oil spill. *Proceedings of the National Academy of Sciences of the United States of America* 109(50):20222–20228.

NAE and NRC (National Academy of Engineering and National Research Council). 2011. *Macondo Well Deepwater Horizon Blowout: Lessons for Improving Offshore Drilling Safety.* Washington, DC: The National Academies Press.

National Academy of Sciences and Royal Society. 2014. *Climate Change: Evidence & Causes.* Washington, DC: The National Academies Press.

National Science Board. 2005. *Report of the National Science Board on the National Science Foundation's Merit Review System.* Arlington, VA: National Science Foundation. Available at http://www.nsf.gov/nsb/documents/2005/nsb05119.pdf (accessed September 2, 2014).

NRC (National Research Council). 2010. *Advancing the Science of Climate Change.* Washington, DC: The National Academies Press.

NRC. 2011. *America's Climate Choices.* Washington, DC: The National Academies Press.

Priest, T. 2013. History of offshore oil and gas in the Gulf of Mexico. Presentation at the Gulf Research Program Advisory Group Meeting, New Orleans, July 23–25.

Sidlauskas, B., G. Ganapathy, E. Hazkani-Covo, K. P. Jenkins, H. Lapp, L. W. McCall, S. Price, R. Scherle, P. A. Spaeth, and D. M. Kidd. 2009. Linking big: The continuing promise of evolutionary synthesis. *Evolution* 64(4):871–880.

Sung, N. S., W. F. Crowley, Jr., M. Genel, P. Salber, L. Sandy, L. M. Sherwood, S. B. Johnson, V. Catanese, H. Tilson, K. Getz, E. L. Larson, D. Scheinberg, E. A. Reece, H. Slavkin, A. Dobs, J. Grebb, R. A. Martinez, A. Korn, and D. Rimoin. 2003. Central challenges facing the clinical research enterprise. *JAMA* 289(10):1278–1287.

Weiss, J. L., J. T. Overpeck, and B. Strauss. 2011. Implications of recent sea level rise science for low-elevation areas in coastal cities of the conterminous U.S.A. *Climatic Change* 105:635–645.

APPENDIXES

Appendix A   Advisory Group Biographies

Barbara A. Schaal, *Chair* **(NAS)** is dean of the Faculty of Arts & Sciences, Washington University, St. Louis. She is chair of the Division on Earth and Life Studies at the NRC and is on President Obama's Council of Advisors for Science and Technology. Her research focuses on the evolutionary genomics of rice. She received a B.S. from the University of Illinois, Chicago, and a Ph.D. from Yale University, both in biology. She has been president of the Botanical Society of America and the Society for the Study of Evolution and vice president of the NAS. She is an elected member of the NAS and the American Academy of Arts and Sciences, and was appointed a U.S. science envoy by former Secretary of State Hillary Clinton.

Donald F. Boesch is president of the University of Maryland Center for Environmental Science. During the 1980s he was executive director of the Louisiana Universities Marine Consortium. A biological oceanographer, he has conducted research in coastal and continental shelf environments along the Atlantic and Gulf coasts, Australia, and China. Dr. Boesch has long been active in extending knowledge to environmental and resource management at regional, national, and international levels. He was a member of the NRC Committee on America's Climate Choices, chair of the NRC Ocean Studies Board, and one of seven members of the National Commission on the BP *Deepwater Horizon* Oil Spill and Offshore Drilling. Dr. Boesch received his B.S. from Tulane University and Ph.D. from the College of William & Mary.

Robert S. Carney is a professor emeritus in the Department of Oceanography and Coastal Sciences of Louisiana State University. He is a biological oceanographer interested in deep-ocean ecology and human-impacted systems. His research activities in the Gulf of Mexico include 28 years at Louisiana State University (LSU). He has served as director of LSU's Coastal Ecology Institute and founding director of the Coastal Marine Institute, a multidisciplinary cooperative with BOEM. Interaction with the regulatory agencies and offshore industry has included funding via BOEM, NOAA, and consulting. Prior to moving to the Gulf Coast he served as program director of Biological Oceanography at the National Science Foundation.

Stephen R. Carpenter (NAS) is the S.A. Forbes Professor of Zoology and director of the Center for Limnology at the University of Wisconsin–Madison. His research focuses on biogeochemistry and food webs in lakes, ecosystem services, scenario planning for environmental change, and resilience of social–ecological systems. He is on the boards of the Stockholm Resilience Center and the South American Institute for Resilience and Sustainability.

Cortis K. Cooper is a Chevron Fellow, 1 of 25 elite scientist and engineers in the company. His primary technical efforts at Chevron focus on better understanding oil spills and physical oceanographic processes that affect the operation and design of offshore facilities worldwide. To improve tools and databases, he has initiated and led many important cooperative research efforts such as a 32-company consortium to better quantify the risks during the transportation of major offshore facilities; a 24-company consortium that investigated the fate of oil and gas from deepwater blowouts; an 18-company consortium focused on understanding strong physical oceanographic phenomena that can affect offshore facilities and operations; and working groups on the use of subsea dispersants. Dr. Cooper has served on six NAS committees and boards, the board of the Gulf of Mexico Coastal Ocean Observing System Regional Association, and several federal advisory committees.

Courtney Cowart is a scholar in the fields of ascetical theology and American church history. In 2013, she was appointed associate dean and director of the University of the South's School of Theology Programs Center. Shortly after completing her Th.D., and while serving Trinity Grants in New York, Dr. Cowart played a leading role in the recovery ministry at Trinity Church's St. Paul's Chapel following the terrorist attacks on September 11, 2001. Four years later, she was deployed to New Orleans in the days following Hurricane Katrina, where she served as the co-director of the Office of Disaster Response for the Episcopal Bishop of Louisiana. In this role, Dr. Cowart led major collaborations with other nonprofits, FEMA, and other agencies. Her on-the-ground post-Katrina work through the Episcopal Church brought a unique perspective on the communities of the Gulf region to the Advisory Group. She received a Th.D. from the General Theological Seminary of the Episcopal Church of New York in 2001.

Robert A. Duce is University Distinguished Professor Emeritus of Oceanography and Atmospheric Sciences and retired dean, College of Geosciences, at Texas A&M University. He was also dean, Graduate School of Oceanography, University of Rhode Island. He was awarded the Rosenstiel Award for his research in atmospheric and marine chemistry and has more than 300 scientific contributions. He has served on the NRC Board on Atmospheric Sciences and Climate, chaired several NRC committees, and now chairs the NRC Ocean Studies Board. He is a fellow of the American Geophysical Union, American Meteorological Society, American Association for the Advancement of Science, and The Oceanography Society (TOS), and is a member of the Ocean Research Advisory Panel. He has been president of TOS, the International Association of Meteorology and Atmospheric Sciences, the Scientific Committee on Oceanic Research, and the International Commission on Atmospheric Chemistry and Global Pollution, and chair of the United Nations Joint Group of Experts on the Scientific Aspects of Marine Environmental Protection (GESAMP).

Deborah L. Estrin (NAE) is a professor of computer science at Cornell NYC and a professor of public health at Weill Cornell Medical College. She is also a co-founder of the nonprofit startup, Open mHealth. She was previously professor of computer science at the University of California, Los Angeles, and the founding director of the National Science Foundation–funded Center for Embedded Networked Sensing. Dr. Estrin is a pioneer in networked sensing, which uses mobile and wireless systems to collect and analyze real-time data about the physical world and the people who occupy it. Her current focus is on mobile health, leveraging the programmability, proximity, and pervasiveness of mobile devices and the cloud for health management. She also has worked on K–12 education, spearheading a groundbreaking internship program for Los Angeles high school students in mobile technologies and data. Dr. Estrin received her Ph.D. in computer science from the Massachusetts Institute of Technology in 1985.

Christopher B. Field (NAS) is the founding director of the Carnegie Institution's Department of Global Ecology, Melvin and Joan Lane Professor for Interdisciplinary Environmental Studies at Stanford University, and faculty director of Stanford's Jasper Ridge Biological Preserve. Dr. Field's research emphasizes impacts of climate change, from the molecular to the global scale. He has been deeply involved with national- and international-scale efforts to advance science and assessment related to global ecology and climate change. He is co-chair of Working Group II of the IPCC, which is currently working on the IPCC Fifth Assessment Report, scheduled for release in 2014. Dr. Field is a recipient of a Heinz Award and is a member of the NAS. He is a fellow of the American Academy of Arts and Sciences, the American Association for the Advancement of Science, and the Ecological Society of America. Dr. Field received his Ph.D. from Stanford in 1981.

Gerardo Gold-Bouchot is a professor at the Marine Resources Department of the Center for Research and Advanced Studies at Merida (Cinvestav Merida), where he has been director of the campus and chairman of the department. He is the coordinator of the Gulf of Mexico Large Marine Ecosystem Project, a U.S.–Mexico project funded by the Global Environment Facility through the United Nations Industrial Development Organization. He is a member of the GESAMP and a member of a number of other international groups of experts.

Lynn R. Goldman (IOM) is dean of the School of Public Health at George Washington University. From 1999 to 2010, she was a professor of environmental health sciences at the Johns Hopkins Bloomberg School of Public Health as well as principal investigator for the Johns Hopkins National Children's Study Center and dual principal investigator for the National Center of Excellence for the Study of Preparedness and Catastrophic Event Response. She was assistant administrator for the EPA Office of Chemical Safety and Pollution Prevention (1993–1998). Dr. Goldman has conducted public health investigations on pesticides, childhood lead poisoning, hazardous waste, and other environmental hazards. She earned her M.D. from the University of California, San Francisco, M.S. in health and medical sciences from the University of California, Berkeley, M.P.H. from the Johns Hopkins Bloomberg School of Public Health, and B.S. from the University of California, Berkeley.

Bernard D. Goldstein (IOM) is emeritus professor and former dean of the University of Pittsburgh Graduate School of Public Health. His NAS activities include chairing the Committee on Considerations for the Future of Animal Science Research and the Committee on Sustainability at the EPA. He has been a member of the IOM Environmental Health Science Research and Training Roundtable and of the NRC Committee on Risk Management and Governance of Shale Gas. He also serves on numerous committees related to shale gas and energy issues including for the Canadian Council of Academies. He was EPA assistant administrator for research and development (1983–1985). His Gulf-related activities include the IOM's Workshop Assessing the Health Effects of the Gulf of Mexico Oil Spill, the Coordinating Committee of the Gulf Region Health Outreach Program, and the GoMRI Public Health Working Group.

Thomas O. Hunter retired in July 2010 as president and laboratories director of Sandia National Laboratories. Dr. Hunter joined Sandia in 1967 and became president in 2005. His responsibilities included managing the laboratories' $2.3 billion annual budget and 8,700 employees. He led programs in areas of national security, nuclear nonproliferation, energy development, environmental management, and the U.S. R&D Enterprise. He served as leader, with Dr. Steven Chu, of the government's science team for the *DWH* oil spill and chaired the Ocean Energy Safety Advisory Committee for the Department of the Interior. He chairs the Leadership Development Advisory Committee for the University of Florida College of Engineering. He has degrees from the University of Florida and the University of New Mexico and a Ph.D in nuclear engineering from the University of Wisconsin–Madison.

Shirley Ann Jackson (NAE) is president of Rensselaer Polytechnic Institute, in Troy, New York, and Hartford, Connecticut. Dr. Jackson has held senior leadership positions in government, industry, research, and academe. A theoretical physicist, she was chairman of the U.S. Nuclear Regulatory Commission (1995–1999). She serves on the President's Council of Advisors on Science and Technology. Her research and policy focus includes energy security and the national capacity for innovation. She is an International Fellow of the British Royal Academy of Engineering, a member of the NAE, a fellow of the American Association for the Advancement of Science, a regent of the Smithsonian Institution, and a member several other prestigious scientific and policy organizations. She holds an S.B. in physics and a Ph.D. in theoretical elementary particle physics, both from the Massachusetts Institute of Technology.

Ashanti Johnson is the executive director of the Institute for Broadening Participation and the assistant vice provost for Faculty Recruitment and associate professor of earth and environmental science at the University of Texas at Arlington. Her areas of research specialization include aquatic radiogeochemistry, professional development of students and early-career professionals, and science, technology, engineering, and mathematics (STEM) diversity-focused initiatives. Dr. Johnson has received numerous honors and awards. In 2010 she received a Presidential Award for Excellence in Science, Mathematics and Engineering Mentoring at the White House in recognition of her professional development and diversity-related activities and was recognized by TheGrio.com, an NBC product, as 1 of 100 History Makers in the Making. She was profiled in the *Black Enterprise Magazine* March 2011 Issue's "Women in STEM" feature story. Dr. Johnson received her Ph.D. (1999) in oceanography from Texas A&M University, College Station.

David M. Karl (NAS) is a microbial biologist and oceanographer in the School of Ocean and Earth Science and Technology, University of Hawaii, whose research interests include the ecology of microorganisms, microbiological and biogeochemical oceanography, and oceanic productivity. Dr. Karl is director of and was instrumental in establishing the University of Hawaii's Center for Microbial Oceanography for Research and Education, a National Science Foundation–supported Science and Technology Center established to facilitate a more comprehensive understanding of the diverse assemblages of microorganisms in the sea. Dr. Karl has served on NRC committees focused on planning the International Polar Year, stewardship while exploring subglacial lakes in the Antarctic, and reviewing the North Pacific Research Board (Alaska). In the course of his career, Dr. Karl has spent more than 3 full years at sea, including 23 expeditions to Antarctica.

Molly McCammon serves as executive director of the Alaska Ocean Observing System, a coalition of government, academic, and private partners working to integrate ocean and coastal data to improve stakeholder decision making. Prior to this, she served for 10 years as executive director of the *Exxon Valdez* Oil Spill Trustee Council, managing the billion-dollar restoration program following the 1989 oil spill in Alaska. Her leadership positions include current treasurer and past chair of the Integrated Ocean Observing System Association, member of the federal Ocean Research Advisory Panel, Municipality of Anchorage representative on the Cook Inlet Regional Citizens Advisory Council, a public oversight group for Cook Inlet Alaska oil and gas activities, and past member of the NAS Polar Research Board.

Linda A. McCauley (IOM) is professor and dean of Emory University's Nell Hodgson Woodruff School of Nursing. Dr. McCauley has special expertise in the design of epidemiological investigations of environmental hazards and is nationally recognized for her expertise in occupational and environmental health nursing. Her work aims to identify culturally appropriate interventions to decrease the impact of environmental and occupational health hazards in vulnerable populations, including workers and young children. Dr. McCauley was previously the associate dean for research and the Nightingale Professor in Nursing at the University of Pennsylvania School of Nursing. She received a Ph.D. in environmental health and epidemiology from the University of Cincinnati. A member of the IOM, she is active on the Environmental Health Roundtable and the Board on Population Health and Public Health Practice. She currently serves as a member of the National Advisory Environmental Health Sciences Council.

J. Steven Picou is the director of the Coastal Resource & Resiliency Center and professor of sociology at the University of South Alabama. He has published more than 150 peer-reviewed articles, book chapters, and research monographs in the areas of environmental sociology, disasters, and sociological practice. In 2001, Dr. Picou received the Distinguished Contribution Award from the American Sociological Association (ASA), and in 2008 was the recipient of the William Foote Whyte Distinguished Career Award given by the ASA Section on Sociological Practice and Public Sociology. His public sociology activities include the preparation of an amicus brief for the U.S. Supreme Court (2008) on the *Exxon Valdez* disaster and he has co-edited and contributed to two books—*The Exxon Valdez Disaster* (1997) and *The Sociology of Katrina* (2010).

Eduardo Salas is trustee chair and Pegasus Professor of Psychology at the University of Central Florida. He earned a Ph.D. in industrial/organizational psychology at Old Dominion University (1984), and has since co-authored more than 400 journal articles and book chapters as well as edited 25 books on topics such as teamwork, team training, safety, team leadership, expertise, minimizing human error, stress, and decision making. Dr. Salas is a fellow of the American Psychological Association and the Human Factors and Ergonomics Society (HFES), past president of the Society for Industrial and Organizational Psychology and the HFES, and a recipient of two life achievement awards for his work on teams and training.

Kerry Michael St. Pé is the retired executive director of the Barataria-Terrebonne National Estuary Program (1997 to July 2014), a nationally recognized effort dedicated to preserving and restoring the 4.2-million-acre area between the Mississippi and Atchafalaya Rivers in southeast Louisiana. Prior to 1997, Dr. St. Pé worked for 25 years as a field biologist for the Water Pollution Control Division of the Louisiana Department of Environmental Quality. His wetland restoration work has been featured in the book *Bayou Farewell, the Rich Life and Tragic Death of Louisiana's Cajun Coast* and in the PBS documentary *Washing Away: Losing Louisiana,* and the Louisiana Public Broadcasting documentary *Turning the Tide.* In May 2010, Dr. St. Pé was awarded an Honorary Doctorate of Science by Nicholls State University. He was presented with the James Lynn Powell Award in March 2012 by the Nicholls Alumni Federation, the highest honor awarded to an alumnus of the university.

Arnold F. Stancell (NAE) is Turner Professor of Chemical Engineering, Emeritus, at Georgia Tech and has been appointed by President Obama to the National Science Board (2010–2014). He worked in industry for 31 years with Mobil Oil, starting in research (granted 11 patents), and then progressing to senior business management positions as vice president of exploration and production for the United States and subsequently for Europe, the Middle East, and Australia. He led the development of the now $70 billion ExxonMobil–Qatar joint venture in liquefied natural gas to produce and supply natural gas to markets worldwide. He was a member of a group of six NAE members who consulted for the U.S. government shortly after the BP oil spill to advise on near-term steps for improved offshore drilling safety, which were announced by President Obama on May 28, 2010.

LaDon Swann is the director of the Mississippi-Alabama Sea Grant Consortium and director of Auburn University's Marine Program. He received B.S. and M.S. degrees from Tennessee Technological University and a Ph.D. from Purdue University. Dr. Swann has more than 29 years of experience in implementing practical solutions to coastal and Great Lakes issues through competitive research, graduate student training, and extension and outreach and K–12 education. He is actively involved with how to maximize the role of boundary organizations in translational research. He is the president of the National Sea Grant Association and a member of the Ocean Research Advisory Panel. Dr. Swann is a past president of the U.S. Aquaculture Association and a former Peace Corps Volunteer.

James W. Ziglar is senior counsel at VanNess Feldman, LLP, and principal of The Ziglar Group. He has 50 years of experience in law, investment banking, corporate management, education, and public policy. In addition to his 33 years of experience in the private sector, Mr. Ziglar has served in the federal government as Assistant Secretary of the Interior for Water and Science, sergeant at arms of the U.S. Senate, commissioner of the Immigration and Naturalization Service, law clerk to U.S. Supreme Court Justice Harry A. Blackmun, and a staff aide in the U.S. Senate. He was a Distinguished Visiting Professor of Law at the George Washington University Law School, and a resident fellow at the Harvard Kennedy School of Government Institute of Politics. Mr. Ziglar serves on a number of boards, including the Water Science and Technology Board of the Division of Earth and Life Studies and the Board of Councilors of the Radiation Effects Research Foundation.

Mark D. Zoback (NAE) is the Benjamin M. Page Professor of Geophysics at Stanford University. He is also co-director of the Stanford Rock Physics and Borehole Geophysics Project, an industrial consortium. He received a Ph.D. in geophysics from Stanford University. Dr. Zoback conducts research on in situ stress, fault mechanics, and reservoir geomechanics. He is the author of a textbook titled *Reservoir Geomechanics* and was co–principal investigator of the San Andreas Fault Observatory at Depth, the scientific drilling project that drilled and sampled the San Andreas Fault at 3-km depth. Dr. Zoback served as a member of the NRC committee that produced the report *Macondo Well Deepwater Horizon Blowout: Lessons for Improving Offshore Drilling Safety*. He recently served on the NAE committee investigating the *DWH* oil spill and the Secretary of Energy's committee on shale gas development and environmental protection.

Appendix B Charge to the Advisory Group

The Gulf Research Program Advisory Group is charged to develop an overarching vision that guides the Program as it starts. The group was asked to identify strategic directions, mission, and objectives to ensure that the Program has a significant, long-term impact and to identify how the Program operates (e.g., the balance among the three objectives of monitoring, research and development, and education and training in the three realms of oil system safety, environmental resources, and human health) and approaches to implementation, partnerships, and leveraging support. Specifically, the Advisory Group was charged to:

- Lead a series of meetings to be held in Washington, DC, the Gulf region, and virtually to help us understand needs and opportunities in the three areas of our mission.
- Assist in outreach and coordination with other organizations working in the Gulf region, including help in analysis of the mission, strengths, and weaknesses of related external programs and analysis of our mission and comparative strengths and weaknesses.
- Provide the intellectual leadership to guide the writing of a strategic plan for the program, including creation of the Program's mission, goals, and objectives and discussion of how the program will do its work in the first 5 years.
- Identify the strategic actions needed to implement the Program (short, medium, and long term), including, if possible, preliminary thinking about metrics to measure Program impact and data policy.

Appendix C   Outreach Activities

During its tenure, the Gulf Research Program Advisory Group participated in numerous interactions to inform program planning. In addition to outreach meetings organized by the Advisory Group, Program representatives also attended numerous conferences, workshops, and meetings to engage with stakeholders. These outreach activities included conversations with state environmental agency representatives, academic researchers, industry researchers and safety experts, community organizations, federal researchers in the Gulf region, and many others. The NAS, the Gulf Research Program's Advisory Board members, and the staff deeply appreciate the contributions from the many people we met over the past year.

In 2014, these initial engagement opportunities are being followed by three in-depth workshops to explore issues in more depth, Advisory Board meetings, and other community interactions. Over time, the Program will continue to use a variety of mechanisms to engage with stakeholders, learn about relevant activities, and build opportunities for partnerships.

Outreach activities in 2013 and 2014 included

MEETING	LOCATION	DATE
Offshore Operators Committee Meeting	Woodlands, TX	June 5, 2013
Gulf of Mexico Alliance (GOMA) All Hands Meeting	Tampa, FL	June 24–27, 2013
Gulf Research Program Advisory Group Meeting	New Orleans, LA	July 24–25, 2013
Ocean Research Advisory Panel Meeting	Arlington, VA	Aug. 21–22, 2013
Gulf Research Program Advisory Group Meeting	Washington, DC	Aug. 29–30, 2013
National Association of Black Geoscientists Meeting	Houston, TX	Sept. 4–7, 2013
Gulf Research Program Outreach Meeting	Mobile, AL	Sept. 18, 2013
Gulf Research Program Outreach Meeting	Thibodaux, LA	Sept. 25–26, 2013
Society of Environmental Journalists Meeting	Chattanooga, TN	Oct. 2–6, 2013
Gulf Research Program Virtual Listening Meeting	Virtual	Oct. 10, 2013
National Institute of Environmental Health Sciences Meeting	Research Triangle Park, NC	Oct. 22–23, 2013
Gulf Research Program Outreach Meeting	Tallahassee, FL	Oct. 30–31, 2013
Gulf Research Program Outreach Meeting	Long Beach, MS	Nov. 12–13, 2013
Gulf Research Program Outreach Meeting	Austin, TX	Nov. 21–22, 2013
Gulf Research Program Virtual Listening Meeting	Virtual	Dec. 5, 2013

American Geophysical Union Meeting	San Francisco, CA	Dec. 9–13, 2013
Interagency Coordinating Committee on Oil Pollution Research	Washington, DC	Dec. 11, 2013
2014 Gulf of Mexico Oil Spill & Ecosystem Conference	Mobile, AL	Jan. 26–29, 2014
Gulf Research Program Advisory Group Meeting	Houston, TX	Feb. 11–12, 2014
GOMA/National Oceanic and Atmospheric Administration Environmental Monitoring Workshop	Stennis Space Center, MS	Mar. 6, 2014
Summit 2014: State of the Gulf of Mexico	Houston, TX	Mar. 24–27, 2014
Spitfire NGO Gulf Science Working Group	Washington, DC	Apr. 12, 2013
2014 International Oil Spill Conference	Savannah, GA	May 5–8, 2014
Bureau of Ocean Energy Management Outer Continental Shelf Scientific Committee Meeting	Reston, VA	May 13–15, 2013
Gulf Research Program Education & Training Opportunity Analysis Workshop	Tampa, FL	June 9–10, 2014
Gulf Research Program Advisory Group Meeting	Tampa, FL	June 11–12, 2014
Conference on Ecological and Ecosystem Restoration	New Orleans, LA	July 29–30, 2014
Gulf Research Program Environmental Monitoring Opportunity Analysis Workshop	New Orleans, LA	Sept. 3–4, 2014
Gulf Research Program Community Resilience and Health Opportunity Analysis Workshop	New Orleans, LA	Sept. 22–23, 2014

Appendix D   Other Funding Programs

ORGANIZATION	FUNDING SOURCE	AMOUNT
Deepwater Horizon Oil Spill Trust, managed by the Gulf Coast Claims Facility (GCCF)	**BP**	**$20 billion**, $5 billion per year paid by BP
GoMRI (Gulf of Mexico Research Initiative)	**BP**	**$500 million** to be disbursed over 10 years
GRHOP (Gulf Region Health Outreach Program)	**BP** as part of the *Deepwater Horizon* Medical Benefits Class Action Settlement	**$105 million** to be paid over 5 years
NAS (National Academy of Sciences) Gulf Research Program	**BP** ($350 million) and **Transocean** ($150 million)	**$500 million** received from 2013 to 2018, and disbursed over 30 years
NAWCF (North American Wetlands Conservation Fund) managed by the U.S. Fish and Wildlife Service (FWS), the North American Wetlands Conservation Council (NAWCC), and the Migratory Bird Conservation Commission (MBCC)	**BP** fine for violations of the Migratory Bird Treaty Act	**$100 million** received from 2014 to 2019, and disbursed: $20 million within 60 days of sentencing (on January 29, 2013), $20 million within 1 year, $20 million within 2 years, $12 million within 3 years, $12 million within 4 years, and $16 million within 5 years
NFWF (National Fish and Wildlife Foundation)	**BP** ($2.394 billion) and **Transocean** ($150 million) from criminal settlements	**$2.544 billion** received from 2013 to 2018 and disbursed over 5 years
NIEHS (National Institute of Environmental Health Studies) **DWH Research Consortium**	**NIH**	**$25.2 million** over 5 years to the following universities and their community partners: University of Florida; Louisiana State University Health Sciences Center, New Orleans; Tulane University; and The University of Texas Medical Branch at Galveston
NRDA (National Resource Damage Assessment)	**BP**	**$1 billion** voluntarily being paid (currently in progress) by BP to fund certain restoration projects before the NRDA was complete; currently being disbursed
OSLTF (Oil Spill Liability Trust Fund) managed by the U.S. Coast Guard National Pollution Funds Center	**MOEX** ($45 million from civil penalties); **Transocean** ($100 million from criminal penalties, $200 million from civil penalties); **BP** ($1.15 billion from criminal penalties)	**$1.495 billion** received on various schedules from 2012 to 2018
RTF (Restoration Trust Fund) created by the RESTORE Act and managed by the Department of the Treasury	**Transocean** ($800 million) and **BP** (amount TBD) for civil Clean Water Act liabilities	**$800 million with additional funds expected (as court cases are resolved).** Initial funds received from 2013 to 2015. Disbursement schedule varies by process

PURPOSE	SPECIFIC ALLOCATIONS
Compensate for natural resource damages, state and local response costs, and individual financial damage	
Fund research projects and consortia to understand, respond to, and mitigate the impacts of petroleum pollution and related stressors of the marine and coastal ecosystems, with an emphasis on the Gulf of Mexico	
Inform residents of the Gulf region about their own health and facilitate access, now and in the future, to skilled frontline health care providers supported by networks of specialists knowledgeable in addressing physical, behavioral, and mental health needs	$50 million to the Primary Care Capacity Project; $36 million to the Mental and Behavioral Health Capacity Project; $4 million to the Community Health Workers Training Project; and $15 million to the Environmental Health Capacity and Literacy Project
Establish a new research program focused on human health and environmental protection in the Gulf of Mexico and on the U.S. outer continental shelf, including issues relating to offshore oil drilling and hydrocarbon production and transportation	For work in three program areas: oil system safety, human health, and environmental resources using three approaches: research and development, education and training, and environmental monitoring. Allocation among program areas and approaches were not specified
Fund wetlands restoration and conservation projects located in states bordering the Gulf of Mexico or otherwise designated to benefit migratory bird species and other wildlife and habitat affected by the oil spill	
Remedy harm and eliminate or reduce the risk of future harm to Gulf Coast natural resources that were adversely affected by the *DWH* explosion and oil spill	$1.272 billion for barrier island and river diversion projects in LA; $356 million for natural resource projects in each of AL, FL, and MS; $203 million for projects in TX
Create community–university partnerships to examine the long-term impact from the oil spill on the health of Gulf Coast residents and communities. This NIEHS research initiative and other related programs help communities and institutions in the Gulf and around the country understand how to be prepared for disasters and limit negative health effects related to disasters	
Restore natural resources impacted by the spill to the condition they would have been in had the spill not occurred	NOAA and the Department of the Interior (DOI) will each receive $100 million for projects to restore federal trust resources. The trustees for each Gulf State will receive $100 million; $300 million will be used for restoration projects that the state trustees suggest, and that NOAA and DOI select
Fund federal agencies to administer the Oil Pollution Act (OPA), respond to future oil spills, and support research and development	$50 million for the Emergency Fund, used for spill response and to initiate natural resource damage assessments. The rest is for the Principal Fund used to compensate those harmed by an oil spill when responsible parties cannot pay and, when appropriated by Congress to cover the costs of administering provisions of the OPA
Varies with the process, but generally for restoration and protection of the natural resources, ecosystems, and economies of the Gulf Coast	35% of the fund goes directly to the five Gulf States in equal shares; 30% goes to a regional Gulf Coast Ecosystem Restoration Council; 30% goes to the five Gulf States based on a formula that considers their respective disturbance from the *DWH* oil spill; 2.5% will support a NOAA-led Gulf Coast Ecosystem Restoration Science, Observation, Monitoring, and Technology program; 2.5% will sustain a competitive grant program to establish Centers of Excellence to conduct Gulf Coast research

SOURCES: Environmental Law Institute and Tulane Institute on Water Resources Law and Policy (2014); The BP Claims Fund website (accessed May 2014); Gulf of Mexico Research Initiative website (accessed May 2014); National Institute of Environmental Health Sciences website (accessed May 2014); and Louisiana Public Health Institute Primary Care Capacity Project, Gulf Region Health Outreach Program website (accessed May 2014).

CREDITS

Cover: A view of the Gulf of Mexico from space. Source: SeaWiFS Project, NASA/GSFC, ORBIMAGE.
p. i: The Gulf of Mexico extends toward the horizon. Source: ©iStock.
p. iv: A satellite image of the Mississippi River delta. Source: Image courtesy NASA/GSFC/METI/ERSDAC/JAROS, and U.S./Japan.
p. 10: People enjoy the beach on South Padre Island, Texas. Source: ©iStock.
p. 14: A hiker explores Louisiana's Manchac Swamp. Source: ©iStock.
p. 15: Seagulls are drawn to a shrimp boat in the Gulf of Mexico. Source: ©iStock.
p. 18: Source: ©iStock.
p. 20: A hawksbill sea turtle swims past a diver in Flower Gardens National Marine Sanctuary in the Gulf of Mexico. Source: ©iStock.
p. 22: Water snakes through the Grand Bay National Estuarine Research Reserve in Mississippi. Source: NOAA.
p. 28: A team in Barataria Bay, Louisiana, evaluates oiled marshes. Source: NOAA.
p. 29: A pelican eyes the camera. Source: ©iStock.
p. 30: Brain coral off of the Florida coast. Source: NOAA.
p. 32: Experts aboard NOAA's research ship *Okeanos Explorer* perform maintenance on remotely operated vehicles during a research cruise in the Gulf of Mexico. Source: Image courtesy of NOAA *Okeanos Explorer* Program, Gulf of Mexico 2014 Expedition.
p. 35: A pair of biological technicians assess sediment gathered from the sea floor in the Gulf of Mexico. Source: Image courtesy of Lophelia II 2012 Expedition, NOAA-OER/BOEM.
p. 38: Offshore oil platforms and a supply ship in the Gulf of Mexico. Source: ©iStock.
p. 40: A lone oystercatcher spreads its wings in shallow water off the Florida coast. Source: ©iStock.
p. 44: Limitless Vistas students Calvin Pitcher and Rolnesha Meyers learn to test for dissolved oxygen on water collected from Twin Canals at the S.E. Louisiana Jean Lafitte National Historical Park in the Barataria Preserve, Louisiana. Source: Limitless Vistas.
p. 48: Boom extends into the Gulf of Mexico from Dauphin Island. Source: ©iStock.
p. 51: Permits swim past an underwater camera in Florida's Dry Tortugas. Source: NOAA.